Basic Mathematics
write-in text

by

Walter C. Brown
Professor Emeritus,
Division of Technology
Arizona State University, Tempe

Publisher
The Goodheart-Willcox Company, Inc.
Tinley Park, Illinois

INTRODUCTION

Basic Mathematics will help you improve your skill in math fundamentals.

Basic Mathematics provides skill-building in addition, subtraction, multiplication, and division of whole numbers, mixed numbers, decimals, and fractions. You will learn to convert between decimals and common fractions. The use of the pocket calculator is also covered in this book. Individual lessons will teach you to measure, read a rule, and use a micrometer.

Basic Mathematics is a combination text and workbook. The text tells and shows *how* while the workbook portion provides exercises and drills that will enable you to develop accuracy and speed in the various topics. Practical problems are included to familiarize you with the application of mathematics in various occupations. Spaces are provided in the book for you to work the problems and write the answers. By referring to the self-check answers in the back of the book, you can determine if your answers are correct.

Basic Mathematics is intended for anyone who needs to improve or review mathematics skills. It may be used as a self-help home study course or in formal classes. **Basic Mathematics** will prove useful in the following applications:

- Industrial and apprentice training programs
- Tech-prep curriculums
- Trade and industry programs
- Vocational/applied math classes
- Individuals improving their employability skills
- Joining academic concepts to technical applications

Study the examples, complete the problems, show your work, and check your answers to become successful on the job. This book is important to improving your employability. Enjoy and learn for your future.

About the Author

Dr. Walter C. Brown taught basic mathematics to students in secondary schools and adult education programs. His background was in industrial technical education and he authored several books in the field of drafting and printreading. He taught and held administrative positions in vocational technical programs in high schools and colleges.

Dr. Brown served as a consultant to industry, state, and local school systems, community colleges, and university programs.

CONTENTS

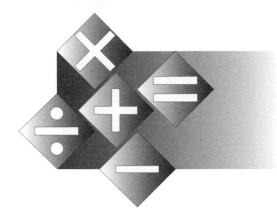

WRITING AND READING NUMBERS

A Number is a figure or a word that indicates a quantity, or sum.

NUMBERS USED

Numbers we use are: 1 (one); 2 (two); 3 (three); 4 (four); 5 (five); 6 (six); 7 (seven); 8 (eight); 9 (nine). It is sometimes necessary to have a number that indicates "not any." The 0 (zero) is used for this purpose. By combining numbers 1 to 9, and the 0, we can indicate any desired quantity.

Example: 4 (four) tools are shown in this drawing.

HOW TO WRITE AND READ NUMBERS

Numbers of one thousand (1,000) or more, are usually separated by commas in groups of three, starting at the right. Omitting the commas is not incorrect, however, when working problems.

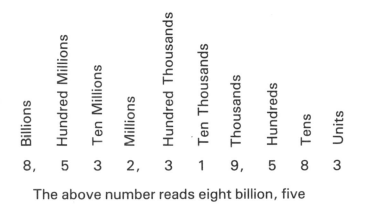

Billions	Hundred Millions	Ten Millions	Millions	Hundred Thousands	Ten Thousands	Thousands	Hundreds	Tens	Units
8,	5	3	2,	3	1	9,	5	8	3

The above number reads eight billion, five hundred thirty-two million, three hundred nineteen thousand, five hundred eighty-three.

Number Reading Examples:

47—Forty-seven.

682—Six hundred eighty-two.

2,499—Two thousand, four hundred ninety-nine.

176,465—One hundred seventy six thousand, four hundred sixty-five.

11,499,678—Eleven million, four hundred ninety-nine thousand, six hundred seventy-eight.

EXERCISE 1-1

Complete the following exercises. Check your answers (page 129).

1. Write 26 in words

_____

2. Write 83 in words

_____

3. Write 478 in words

_____

4. Write 597 in words

_____

5. Write 23,404 in words

_____

6. Write 105,777 in words

_____

_____

7. Write 641,000 in words

_____

_____

8. Write 100,001 in words

_____

_____

9. Write 409,546,223 in words

_____

_____

10. Write 682,051,439 in words

_____

_____

_____

11. Write the numbers for ninety-six

_____

12. Write the numbers for seventy-three

_____

13. Write the numbers for six hundred four

_____

14. Write the numbers for two hundred

_____

15. Write the numbers for eight hundred twenty-three

_____

16. Write the numbers for twenty-three thousand, four hundred twenty-two

_____

17. Write the numbers for eighty-eight thousand, six hundred thirty-three

_____

18. Write the numbers for nine hundred forty-seven thousand, six hundred eighty-two

_____

19. Write the numbers for one hundred one thousand, one hundred one

_____

20. Write the numbers for nine hundred seventy-eight thousand, six hundred thirty-three

_____

21. Write the numbers for eight million, four hundred fourteen thousand, seven hundred ninety-two

_____

22. Write the numbers for three million, two hundred fourteen thousand, six

_____

23. Write the numbers for forty-eight million, thirty-seven thousand

_____

THE METRIC SYSTEM

The inch-pound (customary) measuring system was adopted early in this country for two principal reasons: (1) Most of our trade was with England where the inch-pound system was in use. (2) While the metric system was more uniform in its application to weights and measures, it was not firmly established in any nation including France where it had its beginning.

Today, the International (SI) Metric System is in use in most nations of the world and the United States is moving toward metrication.

This section (Unit 2) will provide you with an understanding of the metric system units of measurement and their prefixes.

In beginning your study, it is well to remember that the metric system is no more than a simple language of measurement. Most of us will use only a small portion of the entire system. This book presents those units that you are most likely to use—length, area, weight, volume, and temperature. The metric system is much simpler than the customary measuring system because it is based on the decimal system just like our money system.

CUSTOMARY UNITS

In the customary system we use the foot as a measure of length. If we want to use units larger than this, we speak of a yard, fathom, rod, or mile. For units smaller than the foot, we use the inch and fractions of the inch. It will be noted that neither the names of the units nor their multiples bear any systematic relationship. The same is true for the measure of weight in the customary units—ounces, pounds, and tons.

METRIC UNITS

The relationship of metric units is more understandable and systematic. Units are related by a divisor or multiplier of 10.

Prefixes of Metric Units

A standard unit has been used in each area of measure and all units, smaller or larger, are identified by special names, called prefixes, which go before the name of the unit. For example, the **Meter** is the standard unit of measure for length and all other units of length hold a definite relationship to the meter as expressed by the prefix. Instead of a comma, a space separates groups of numbers in metric (26 000, not 26,000).

millimeter (mm) is $\dfrac{1}{1000}$ of a meter
(there are 1000 mm in a meter)

centimeter (cm) is $\dfrac{1}{100}$ of a meter
(there are 100 cm in a meter)

decimeter (dm) is $\dfrac{1}{10}$ of a meter
(there are 10 dm in a meter)

meter (m) is 1 meter

dekameter (dam) is 10 meters

hectometer (hm) is 100 meters

kilometer (km) is 1000 meters

Notice that all prefixes are tied to the unit (meter in this case) and have a definite meaning.

That is, 4 millimeters means $\dfrac{4}{1000}$ of a meter

$\dfrac{1}{1000}$	$\dfrac{1}{100}$	$\dfrac{1}{10}$	Unit	10	100	1000
Millimeter	Centimeter	Decimeter	Meter	Dekameter	Hectometer	Kilometer

PREFIX	MEANING	EXAMPLE
MILLI	= thousandths	1 millimeter $= \dfrac{1}{1000}$ meter
CENTI	= hundredths	3 centigrams $= \dfrac{3}{100}$ gram
DECI	= tenths	6 deciliter $= \dfrac{6}{10}$ liter
DEKA	= tens	2 dekameters $= 20$ meters
HECTO	= hundreds	4 hectograms $= 400$ grams
KILO	= thousands	7 kilometers $= 7000$ meters

and 4 kilometers means 4000 meters. These prefixes are subdivisions or multiples of 10 and are used for all units of measure in the metric system. You should learn these prefixes and their meanings now!

METRIC MEASURE OF LENGTH

We have already learned the meter is the standard unit of length in metric. All sub-units and multiple-units are related by a divisor or multiplier of 10 as shown at the top of this page.

The meter is slightly longer than the customary unit of a yard. This is too large of a unit to use for measurements in the machine manufacturing industry. Therefore, the smaller sub-unit millimeter is used. There are approximately 25 millimeters in an inch. It is easier to read and understand the size of a small machine part when it is dimensioned in (a) millimeters rather than (b) meters:

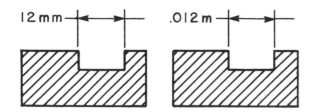

In the construction industry, the centimeter and the meter are used for linear measurements.

The frontage of a residential building lot will likely be given in meters:

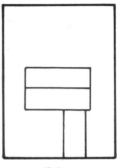

30 m

When a greater distance is referred to, such as the distance between two cities, it usually is given in kilometers rather than meters:

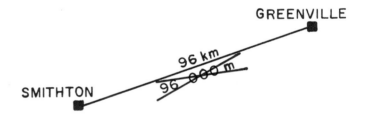

METRIC MEASURE OF MASS (WEIGHT)

The standard metric unit for mass is the **Kilogram** that weighs slightly more than 2 pounds. The same prefixes are used to identify the smaller units. Following are the prefixes more commonly used:

$$1 \text{ milligram (mg)} = \frac{1}{1000} \text{ gram (g)}$$

$$1000 \text{ milligrams} = 1 \text{ gram}$$

$$1 \text{ gram} = \frac{1}{1000} \text{ kilogram (kg)}$$

$$1000 \text{ grams} = 1 \text{ kilogram}$$

When it is necessary to express larger units of weight, the megagram (Mg) is used. The megagram equals 1000 kilograms and is also known as the **Metric ton.**

METRIC MEASURE OF VOLUME

Volume is the cubic measure of a material such as dirt, sand, and gravel. These are usually **measured in Cubic meters** (m^3). When the material to be measured is a fluid, the standard measure is in **Liters,** roughly the size of the customary quart. The liter is the special name given to the cubic decimeter for fluid measures:

millimeter (mℓ) centiliter (cℓ) deciliter (dℓ) liter (ℓ) dekaliter (daℓ) hectoliter (hℓ) kiloliter (kℓ)

EXERCISE 2-1

The following questions and activities will assist you in understanding the metric system. Place your responses in the column at the right. Check your answers on page 129.

1. There are _____ millimeters in a meter.

    1. _____

2. There are _____ centimeters in a meter.

    2. _____

3. How many meters are in a kilometer?

    3. _____

4. Write the meaning of the following prefixes:

    a. milli   d. deka

    b. centi   e. hecto

    c. deci   f. kilo

    4. a. _____

        b. _____

        c. _____

        d. _____

        e. _____

        f. _____

5. The standard unit of length in metric is _____.

    5. _____

6. What unit of metric measure of length is used in the

    a. machine manufacturing industry?

    b. construction?

    6. a. _____

        b. _____

7. The standard unit for weight in metric is _____.

    7. _____

8. The kilogram weighs slightly more than _____ pounds.

    8. _____

9. The metric volume unit of measure for material such as sand is _____.

    9. _____

10. Fluid materials are measured in _____ in metric.

    10. _____

11. How many milliliters are contained in a liter?

    11. _____

12. There are _____ liters in a kiloliter.

    12. _____

13. Estimate the following measurements in meters:

    a. Width of classroom.   c. Height of ceiling.

    b. Length of classroom.   d. Height of door.

    13. a. _____

        b. _____

        c. _____

        d. _____

14. Estimate the following in millimeters:

    a. Length of a chalkboard eraser.

    b. Width of a chalkboard eraser.

    c. Height, width, and thickness of this math book.

14. a. _____

    b. _____

    c. H _____ W _____

       T _____

15. Estimate the weight of the following in kilograms:

    a. This math text.

    b. Your own weight.

15. a. _____

    b. _____

16. Estimate the volume of the following in liters:

    a. One gallon of paint.

    b. Gasoline tank on your car.

16. a. _____

    b. _____

Note: For more information on the metric system and how to convert (change) from customary units to metric units, see Unit 19, page 73 and the Appendix, page 116.

Addition is the process of combining or putting together two or more numbers to find out how many there are altogether. The figure we get is called the **Sum**.

The sign used to indicate that numbers are to be added is the **Plus sign**, written (+). Numbers written left to right that are to be added are written 4 + 2 + 2.

To indicate the sum or total quantity obtained when adding we use the (=) that is called an **Equals sign**. Example: 4 + 2 + 2 = 8.

The problem could also be written in a column like this:

```
4
2
2
8 Sum
```

When numbers to be added are written in a column, the units are written under units, tens under tens, thousands under thousands, etc.

Example:

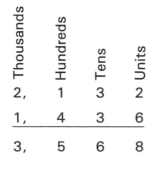

Thousands	Hundreds	Tens	Units
2,	1	3	2
1,	4	3	6
3,	5	6	8

In adding, always start with the column of numbers at the right. When the sum of a column is more than 9, the second number is added to the column at the left. This is called **Carrying**.

Example:

```
  ′            ′ ′
3 1 9        7 8 9
4 3 8        5 7 2
7 5 7        1361
```

Study the examples carefully, to see how the problems were solved, then use the same procedure to solve the following problems.

EXERCISE 3-1

Add the following. Show numbers carried. Check your answers (page 129).

1. 302 515 298	2. 414 379 708
3. 416 329 384 171	4. 115 87 210 468
5. 9372 4765 1019 9988	6. 2075 59 9091 1294
7. 8742 3178 7333	8. 9199 4073 9194

9. 9184
    3777
    4001
    6183
    1672

10. 4188
    6279
    4366
    3555
    1097

11. 2909
    9351
    1287
    6813
    5476

12. 3777
    9368
    8491
    1705
    6917

13. 8877
    2345
    6182
    9009
    1682

14. 8173
    3062
    5155
    8877
    9199

EXERCISE 3-2

Application of addition. Show your work. Check your answers.

1. A salesperson worked the following hours during a five day period: Monday–5, Tuesday–8, Wednesday–6, Thursday–8, and Saturday–4. What were the total hours worked during the week?

2. A machine shop ordered the following amounts of cold rolled steel: 1714 lb., 597 lb., and 1043 lb. How many pounds of steel were ordered?

3. A plumber, in connecting a water tank, used short lengths of pipe as follows: 16", 28", 8", 21", 6", and 32". How many inches of pipe were needed to do the job?

4. An electrician used the following lengths of conduit in wiring a storage shed: 175 cm, 312 cm, 236 cm, 304 cm, and 428 cm. Find the total quantity of conduit used, in centimeters.

5. During a five year period, a worker contributed the following amounts to the company pension fund: $981, $915, $1203, $1209, and $1215. How much was contributed to the pension fund during the five year period?

6. Find the total weight, in kilograms, of the following loads of sand: 9556, 9042, 7983, 10,037, and 9180kg.

7. Part of the job of the head waitress at a small restaurant was to total the amounts deposited in the bank during the week. The deposits were as follows: 1116.83, 1403.18, 1292.53, 982.75, 1056.13, 1968.24, and 1307.27. What is the total amount of the deposits for the seven day period?

Subtraction is the process of withdrawing or taking away. The larger number is called the **Minuend,** the smaller number (number to be withdrawn) is the **Subtrahend,** and the difference between the two numbers is the **Remainder** (or Difference).

The sign that is used to indicate that one number is to be subtracted from the other is the **Minus sign** and is written (−).

Example: 347 − 225 or

```
  347 Minuend
− 225 Subtrahend
  122 Remainder
```

Now let us try an example that is different; one that requires borrowing.

Example: 46 − 27 or

```
 ³⁄4 ¹6 Minuend
 2 7  Subtrahend
 1 9  Remainder
```

Begin with the column at the right (units column). 7 cannot be subtracted from 6, because 7 is greater than 6. Therefore, we must **Borrow** 1 ten or 10 units from the tens column (4) and add this to the 6 units, giving us a total of 16 units. 16 units minus 7 units equals 9 units, as shown. To complete the problem, 3 tens minus 2 tens equals 1 ten.

CHECKING SUBTRACTION PROBLEMS

Checking a problem in subtraction is easy. You simply add the remainder to the subtrahend. If the sum equals the minuend, your subtraction is correct.

Example: 49 − 24

```
Subtract 49          To Check, add 25
         24                          24
         25                          49
```

EXERCISE 4-1

Complete the following exercise. Check each answer.

```
                    Check
1.  46
    13


                    Check
2.  25
    16


                    Check
3.  76
    49


                    Check
4.  946
    214


                    Check
5.  922
    249          .


                    Check
6.  572
     94
```

BASIC MATHEMATICS

7. 974
    712

8. 174
    31

9. 684
    49

10. 422
    321

11. 775
    391

12. 2002
    342

13. 9752
    6315

14. 8967
    4342

15. 20,000
    4,768

16. 7246
    3389

17. 9,460,230
    821,000

18. 65,008
    2,039

19. 410,002
    99,999

20. 93,000,000
    186,000

21. 6240
    0

22. 3976
    3976

23. 0
    0

24. 20,046
    2,004

25. 2743
    108

26. 8765
    8764

27. 9,000
    999

28. 40,098
    3,456

Check

29. 42,111
    2,222

30. 23,736
    16,349

31. 16,432
    15,952

32. 354,927
    115,982

33. 92,000
    186

3. A painting crew used paint from a 50 gallons drum. If the drum was full at the beginning of the day and the painters used 29 gallons, how much was left at the end of the day?

4. An electrician cut a 233 foot length cable from a 1,000 foot roll. How much was left on the roll?

5. Find missing dimension S.

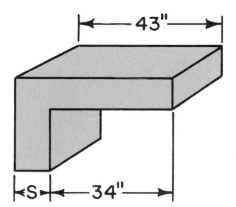

EXERCISE 4-2

Solve these problems. Show your work.

1. A gas meter read 1246 cubic feet on July 5. On August 5, it read 1314 cubic feet Find the number of cubic feet of gas used in that period.

2. A plumbing contractor has 1457 feet of pipe in the warehouse. 239 feet are used on a job. How much pipe is remaining?

6. An automobile trailer and its load weighed 1690 pounds. If the trailer weighed 840 pounds, what was the weight of the load?

7. A contractor had 1940 board feet of studding on hand. Carpenters used 1485 board feet of it in building a house. How many board feet were left?

8. North High School has 123 students enrolled in CAD classes and 27 students enrolled in an electronics class. How many more students are enrolled in CAD classes than in electronics?

9. Find missing dimension A. Dimensions are in millimeters.

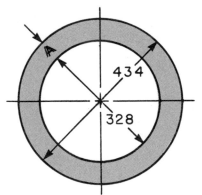

10. A salesperson in a yard-goods store had 288 yards of canvas cloth in stock. She sold 47 yards to a customer for a special job. How many yards were left in stock?

11. A contractor agreed to remove 640 cubic meters of earth for a basement. The crew removed 223 cubic meters the first day. How many cubic meters were left to be removed?

12. The speedometer on an automobile read 49 689 kilometers at the beginning of a trip. At the end of the trip it read 51 214 kilometers. How far did the automobile travel?

13. A machinist has a bar of steel weighing 187 kilograms. If 94 kilograms are used to make die stocks, how many kilograms are left?

14. A masonry contractor ordered 2200 concrete blocks for a construction job. When the project was finished, 46 blocks were remaining. How many blocks were used?

15. A storage bin contains 10,800 bolts. If 3,467 are removed, how many are left?

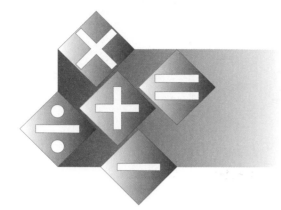

MULTIPLICATION

Multiplication is a short cut method of obtaining the answer to a problem involving the repeated addition of the same number. The sign used to indicate that the problem is to be solved by multiplication is the **Times sign** (×).

Let us assume that we have the problem 4 × 3 (four times three). We could get the answer 12, by adding 4 + 4 + 4 = 12, but there is an easier way by using Multiplication.

Example:

  4 **Multiplicand**
  3 **Multiplier**
12 **Product**

The number that could be added a certain number of times to get the answer (4) is called the **Multiplicand**. The number that shows how many times the 4 is to be added (3) is called the **Multiplier**. The answer (12) is called the **Product**.

Example: Multiply 116 × 12.

```
  116
   12
  232
  116
 1392
```

In multiplication, every number in the multiplicand is multiplied by every number in the multiplier. When the number obtained by multiplying is greater than 9 (as 2 × 6), the second figure is carred over the next column.

CHECKING MULTIPLICATION PROBLEMS

A multiplication problem may be checked by multiplying the numbers in reverse order.

	1	2	3	4	5	6	7	8	9
1	1	2	3	4	5	6	7	8	9
2	2	4	6	8	10	12	14	16	18
3	3	6	9	12	15	18	21	24	27
4	4	8	12	16	20	24	28	32	36
5	5	10	15	20	25	30	35	40	45
6	6	12	18	24	30	36	42	48	54
7	7	14	21	28	35	42	49	56	63
8	8	16	24	32	40	48	56	64	72
9	9	18	27	36	45	54	63	72	81

Multiplication Tables

Example: Multiply 12 × 116.

```
    12
   116
    72
    12
    12
  1392
```

USING MULTIPLICATION TABLES

To multiply quickly and accurately we need to know the multiplication tables. For quick reference and review, it will be helpful to you to make up on a separate sheet of paper, a table like the one shown on the previous page.

EXERCISE 5-1

Study the example, then complete the following exercises. Show your work. Check where indicated.

1.
```
         15        Check
         13          13
         45          15
         15          65
        195          13
                    195
```

2.
```
         25        Check
         23          23
                     25
```

3.
```
         19        Check
         46
```

4.
```
         93        Check
         43
```

5.
```
         57        Check
         72
```

6.
```
         73
         29
```

7.
```
       2312        Check
         23          23
       6936        2312
       4624          46
      53176          23
                     69
                     46
                   53176
```

8.
```
       4031
         40
```

9.
```
      41010
         72
```

10.
```
      49327
         67
```

11.
```
      72130
         43
```

MULTIPLICATION

12.   87750   17.   633
        70   497

13.   214   18.   978
      122   253

14.   743   19.   999
      462   99

15.   300   20.   54876
      204   3086

16.   106
      407

EXERCISE 5-2

Solve the following problems. Show your work.

1. A power hacksaw operates at 140 strokes per minute. How many strokes does it make in 25 minutes?

2. A circular power saw will rip a 10 foot board in 5 seconds. How long will it take to rip fourteen 10 foot boards?

3. A wire-wrapping machine uses 2432 inches of wire in wrapping a coil. How much wire is needed to wrap 15 such coils?

4. A casting weighs 24 kilograms. What would be the weight of 82 of these castings?

5. In a certain assembly plant, a total of 148 workers are employed in assembling electronic components. If each worker uses an average of 6 inches of solder per day, how much solder is used by all of the workers in a day?

6. There are 5280 feet in a mile. How many feet are in 512 miles?

7. The shipping weight of each of 12 electric welders is 355 pounds. What is the total shipping weight?

8. There are 32 boxes of candles in stock. Each box contains 24 candles. What is the total number of candles in the 32 boxes?

9. A cabinet shop has 12 two-foot bar clamps that weigh 7 pounds each; 9 three-foot, weighing 8 pounds each; 14 four-foot weighing 9 pounds each; and 18 six-foot weighing 12 pounds each. What is the total weight of all the bar clamps?

10. An order was placed for 18 machinist vises for the production lab. Eight of the vises weigh 13 kilograms each; six weigh 16 kilograms each; two weigh 21 kilograms each; and two weigh 26 kilograms each. Allowing 14 kilograms for the shipping crate, what is the total shipping weight of the 18 vises and crate?

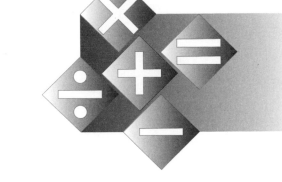

Division is a mathematical procedure by which we find out how many times one number is contained in another number. The sign commonly used to denote division is (÷).

Example: 15 ÷ 5 (fifteen divided by five). Our problem is to find out how many times 5 is contained in 15.

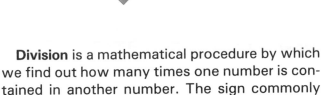

The number to be divided, 15, is called the **Dividend,** the number by which the dividend is divided, 5, is called the **Divisor.** The answer, 3, is called the **Quotient.**

When a number is not contained in another number in an equal number of times, the amount left is called the **Remainder.**

Example: 50 ÷ 6

```
      8
6/50
     48
      2 Remainder
```

CHECKING PROBLEMS

A problem in division may be checked by multiplying the quotient by the divisor and adding in the remainder.

```
  8 Quotient
  6 Divisor
48
  2 Remainder
50
```

EXERCISE 6-1

Complete the following exercises. Show your work. Check your answers by multiplication, where indicated.

1. 65/ 260   Check

2. 14/ 756   Check

3. 37/ 592   Check

4. 18/ 810   Check

Check

5. 13$\overline{)481}$

9. 26$\overline{)12679}$

Check

6. 21$\overline{)378}$

10. 12$\overline{)11952}$

7. 82$\overline{)50773}$

11. 35$\overline{)32306}$

8. 19$\overline{)30782}$

12. 48$\overline{)242883}$

Check

EXERCISE 6-2

13. 394⟌240893

Solve the following problems. Show your work.

1. A piece of lumber is 288 cm long. How many pieces, each 18 cm in length, can be cut from the piece of lumber? Disregard any waste material.

14. 47⟌19918

2. A truck traveled 50,640 miles over a period of one year. If the truck averaged 12 miles per gallon of gasoline, how many gallons of gasoline were used that year?

15. 406⟌34254

3. A truck hauled 6240 kilograms of barbell sets. Each barbell set, with carton, weighed 52 kilograms. How many sets did the truck haul?

16. 88⟌92607

4. A warehouse shipping clerk has 1122 socket wrench sets to be distributed equally to 6 outlet stores. How many should be sent to each store?

5. A machine produces 1512 articles in 24 hours. How many articles does it produce in one hour?

6. An electrician used 2208 meters of electrical cable in wiring 24 apartments. If the apartments were identical, how many meters of cable were needed to wire each of them?

7. A football player carried the ball 32 times for a total gain of 224 yards. What was his average gain per carry?

8. A subcontractor employed 35 men and women for a period of one year. In that year the company paid salaries totaling $924,000. Each employee received the same salary. How much was each paid?

9. Twelve stamping machines produced 312,000 rudder ribs in 40 hours. If each machine operated at the same rate for the full forty hours, how many ribs did each produce per hour?

10. A salesperson makes a profit of $14 each time he/she sells a software program. How many software programs must he/she sell to earn $700?

11. An electric generator produces 1560 watts of power. How many 60 watt light bulbs will it light?

12. A garment worker completes 368 pieces in an 8-hour day. What is the average number of pieces per hour?

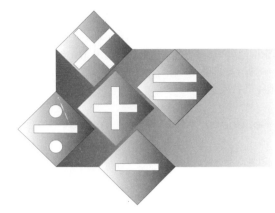

A **Fraction** is one or more parts of the equal parts of a unit or whole number. Fractions are written with one number over the other:

$$\underline{3} \text{ Numerator} \qquad \frac{2}{3} \qquad \frac{11}{16} \qquad \frac{27}{32}$$
$$8 \text{ Denominator}$$

The **Denominator** that indicates the number of parts into which the unit is divided, is written below the line. The **Numerator,** written above the line, indicates the number of parts of the unit with which we are concerned.

In reading a fraction, we always read the top number (numerator) first, then the denominator. The fraction 3/8 would be read Three Eighths.

When the numerator of a fraction such as 3/8 is less than the denominator, the fraction is called a **Proper fraction.** When the numerator is greater than the denominator, as in 9/8 and 17/16, the fraction is called an **Improper fraction.** A **Mixed number** is a number that consists of a whole number and a proper fraction, such as 3 1/2, 5 2/3, 4 7/8.

CHANGING WHOLE NUMBER TO FRACTION

Problem: Change 4 (whole number) to sixths.

$$4 \times \frac{6}{6} = \frac{24}{6}$$

Each whole unit contains 6 sixths. Four units will contain 4 × 6 sixths or 24 sixths.

EXERCISE 7-1

Solve these problems. Show your work.

Change:

1. 49 to sevenths.

2. 40 to eighths.

3. 54 to ninths.

4. 27 to thirds.

5. 12 to fourths.

6. 130 to fifths.

CHANGING MIXED NUMBER (whole number and fraction) TO FRACTION

Problem: Change 3 7/8 (mixed number) to eighths.

Each whole unit contains 8 eighths. Three units will contain 3 × 8 eighths or 24 eighths. Adding the 7/8 we already had, gives us 31 eighths. This would be written $\frac{31}{8}$.

What we have done is to multiply the whole number by the denominator of the fraction, add the numerator, then write the result over the denominator.

EXERCISE 7-2

Solve these problems. Show your work.

Change these mixed numbers to fractions:

1. $4\frac{1}{2}$

2. $8\frac{3}{4}$

3. $19\frac{7}{16}$

4. $7\frac{11}{32}$

5. $6\frac{9}{14}$

6. $5\frac{1}{64}$

REDUCING IMPROPER FRACTION (numerator is greater than denominator) TO WHOLE OR MIXED NUMBER

Problem: Reduce $\frac{18}{3}$ to a whole or mixed number.

Solution: $\frac{18}{3} = 18 \div 3 = 6$. Our answer is 6, a whole number.

Problem: Reduce $\frac{19}{3}$ to a whole or mixed number.

Solution: $\frac{19}{3} = 19 \div 3 = 6\frac{1}{3}$. Our answer is $6\frac{1}{3}$, a mixed number.

EXERCISE 7-3

Solve these problems. Show your work.

Reduce these improper fractions to whole or mixed numbers:

1. $\frac{36}{7}$

2. $\frac{44}{4}$

3. $\frac{23}{5}$

4. $\frac{43}{9}$

5. $\frac{240}{8}$

6. $\frac{191}{6}$

FORM OF FRACTION MAY BE CHANGED WITHOUT CHANGING ITS VALUE

Multiplying or dividing both the numerator and denominator of a fraction by the same number does not change its value.

REDUCING FRACTION TO LOWEST FORM

Problem: Reduce $\frac{4}{6}$ to its lowest form.

Solution: $\frac{4 \div 2}{6 \div 2} = \frac{2}{3}$

Since 2 and 3 contain no common divisor (number that will divide into both numbers), the fraction 2/3 is in its lowest form or denomination. Fractions 4/6 and 2/3 have the same value.

EXERCISE 7-4

Solve these problems. Show your work.

Reduce each of these fractions to its lowest form:

1. $\frac{6}{10}$

2. $\frac{3}{9}$

3. $\frac{6}{64}$

4. $\frac{12}{32}$

5. $\frac{32}{48}$

6. $\frac{76}{152}$

CHANGING FRACTION TO HIGHER FORM OR DENOMINATOR

Problem: Change $\frac{4}{8}$ to 16ths.

Solution: $16 \div 8 = 2$. Multiply both numerator and denominator by 2, like this:

$\frac{4 \times 2}{8 \times 2} = \frac{8}{16}$

Fractions 4/8 and 8/16 have the same value.

EXERCISE 7-5

Solve these problems. Show your work.

Change:

1. 6/8 to 64ths

2. 3/16 to 32nds

3. 11/56 to 168ths

4. 3/7 to 49ths

5. 9/13 to 104ths

6. 31/79 to 237ths

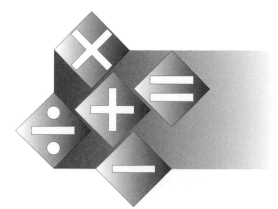

ADDITION OF FRACTIONS, MIXED, WHOLE NUMBERS

When adding fractions, it is necessary that we change all the denominators into what is called the **Lowest Common Denominator** (L.C.D.). That is, we must find the smallest number into which all the denominators can be divided.

FINDING LOWEST COMMON DENOMINATOR

Problem: Add 1/2, 3/4, and 5/8.

Frequently, it is possible to determine the Lowest Common Denominator at a glance. In this case, it is obvious that the denominators 2, 4, and 8, can be divided into 8.

Solution:

$$\frac{1}{2} = \frac{4}{8}$$
$$\frac{3}{4} = \frac{6}{8}$$
$$\frac{5}{8} = \frac{5}{8}$$
$$\frac{15}{8} \text{ or } 1\frac{7}{8}$$

When we have several fractions to be added and the lowest common denominator cannot be determined at a glance, or by simple figuring, we proceed as follows:

Problem: Add 3/8, 9/12, and 6/10.

To find the L.C.D. we write the three denominators in line like this:

$$
\begin{array}{r|ccc}
2 & 8 & 12 & 10 \\
\hline
2 & 4 & 6 & 5 \\
\hline
 & 2 & 3 & 5 \\
\end{array}
$$

Then, we find the smallest number that can be divided into one or more of the denominators. Repeat as many times as you can find two or more divisors.

When a number cannot be divided by the divisor, bring the number (5) down, as shown. Now multiply $2 \times 2 \times 2 \times 3 \times 5 = 120$, the L.C.D. for 8, 12, and 10.

$$\frac{3}{8} = \frac{45}{120}$$
$$\frac{9}{12} = \frac{90}{120}$$
$$\frac{6}{10} = \frac{72}{120}$$
$$\frac{207}{120} = 1\frac{87}{120}$$

EXERCISE 8-1

Add the following. Reduce to lowest terms. Show your work.

1. 3/4 + 3/4 =

2. 1/7 + 0 =

ADDITION OF FRACTIONS, MIXED, WHOLE NUMBERS

3. 5/8 + 3/8 =

4. 2/3 + 7/9 =

5. 5/16 + 7/8 =

6. 3/4 + 3/8 =

7. 9/32 + 17/16 =

8. 2 9/10 + 1/10 =

9. 2/5 + 7/10 =

10. 7/8 + 3/6 =

11. 2 1/7 + 3 2/8 =

12. 4 1/8 + 2 1/4 =

13. 100 1/8 + 99 7/8 =

14. 25 3/4 + 3/12 =

15. 12 3/16 + 6 5/16 =

16. 1/16 + 3/8 + 3/4 =

17. 10/99 + 15/11 + 1/33 =

18. 4/18 + 9/36 + 1/9 =

19. 1 3/20 + 2 3/4 + 5 7/10 + 4/15 =

20. 5 1/2 + 6 2/3 + 1/3 =

21. 85 5/12 + 25 7/18 =

22. 100 + 23 1/5 + 14 2/3 + 1/15 =

23. 1/2 + 3/4 + 7/20 =

24. 239 7/18 + 234 5/54 =

25. 8 4/5 + 6/7 + 4 5/8 =

26. 3 4/9 + 1/2 + 15 3/5 =

27. 4 7/8 + 7/24 + 10 3/4 =

28. 9 4/7 + 15 3/8 + 6 3/4 =

29. 3 2/3 + 9 1/4 + 5 1/2 =

30. 16 1/8 + 39 5/16 + 21 1/4 =

EXERCISE 8-2

Solve these problems. Show your work.

1. An interior wall of a house is made up of 2×4 studs covered on each side with 3/4" wallboard. If the actual width of a 2×4 stud is 3 1/2", what is the total thickness of the wall?

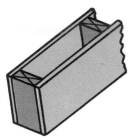

2. A sheet of plywood is made up of five sheets of wood. Two of the sheets are 3/16 inch thick and three of the sheets are 1/8 inch thick. What is the total thickness of the sheet of plywood?

3. Four sheets of metal are stacked in a pile. If the thicknesses of the sheets are 7/32, 7/16, 3/8, and 1/4 inch, what is the total thickness of the stack?

4. A plumber, in making a run of pipe, used three pieces. The first piece was 49 3/8 inches and the other two were each 36 3/4 inches. What was the total length of pipe used?

5. A triangular frame has sides that measure 6 3/4, 8 3/8, and 10 1/2 inches. What is the total length of the three sides?

6. In drilling a well, a worker drilled through 16 1/2 feet of clay, 4 1/4 feet of rock, 20 1/8 feet of gravel, and 40 7/8 feet of sand. How deep was the well?

7. An artist used four pieces of various color poster board to construct a poster. The project required the following widths of poster board to construct: 4 5/8", 7 5/8", 5 3/8", and 9 3/8". How wide was the poster when completed?

8. A fireplace was made by laying a wall of two common bricks 3 3/4 inches in width and one fire brick (facing) 2 1/2 inches in thickness. Between the bricks were two mortar joints, each 5/16 inches thick (two joints). What was the total thickness of the wall?

10. The following lenghts of lumber were cut from a board: 15 3/4, 48, 30 1/2, 23 7/8, and 12 5/8 inches If 1/8 inch of lumber was used in cutting each board, what was the total length used from the board?

9. Five stacks of lumber vary in height as follows: 27 1/2, 32 1/4, 18 5/16, 29 7/8, and 31 3/16 inches. What is the combined height of all 5 stacks?

11. Find the length of the center punch shown below. Dimensions are in inches.

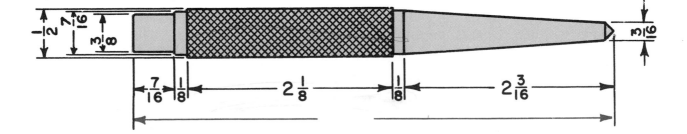

SUBTRACTION OF FRACTIONS, MIXED, WHOLE NUMBERS

When subtracting fractions, with different denominators, the first step is to reduce them to fractions having a **Common (Same) Denominator,** as we did in adding fractions. See page 28. After changing to common denominator, subtract the numerators.

Problem: From 3/4 subtract (take) 1/8.

Solution:

$$\frac{3}{4} = \frac{6}{8}$$
$$-\frac{1}{8} = \frac{1}{8}$$
$$\overline{\qquad \frac{5}{8}}$$

SUBTRACTION OF MIXED NUMBERS

First find the difference between the fractions, then find the difference between the whole numbers.

Problem: From 4 9/16 subtract 2 3/16.

Solution:

$$4\frac{9}{16}$$
$$-2\frac{3}{16}$$
$$\overline{\frac{6}{16} \text{ or } 2\frac{3}{8}}$$

Problem: From 15 1/8 subtract 6 7/8.

Solution:

$$15\frac{1}{8}$$
$$-\ 6\frac{7}{8}$$
$$\overline{\qquad}$$

The denominators are the same, but it is obvious that we cannot take 7/8 from 1/8. We borrow a unit (8/8) from the whole number 15, and add this to the 1/8.

$$1\overset{4}{\cancel{5}}\frac{9}{8}$$
$$-\ 6\frac{7}{8}$$
$$\overline{8\frac{2}{8} \text{ or } 8\frac{1}{4}}$$

EXERCISE 9-1

Subtract the following. Reduce to lowest terms. Show your work.

1. 7/8 − 3/8 =

2. 15/16 − 3/8 =

3. 24/99 − 2/33 =

4. 2 3/4 − 1 1/8 =

5. 3/8 − 3/10 =

6. 23 1/9 − 7 7/10 =

7. 32 − 29 3/8 =

8. 61 16/21 − 5 3/14 =

9. 4 1/8 − 2 7/12 =

10. 90 4/17 − 26 3/4 =

11. 47 7/12 − 9 5/18 =

12. 29 3/4 − 16 29/32 =

13. 5 99/100 − 5 3/5 =

14. 17 3/5 − 7 =

15. 18 3/7 − 8 =

16. 97 − 13 11/12 =

17. 342 − 14 47/131 =

18. 47 3/8 − 47 3/16 =

19. 29 5/8 − 29 5/8 =

26. 8,527 1/4 − 3,750 1/4 =

EXERCISE 9-2

Solve these problems. Show your work.

20. 27 5/13 − 0 =

1. A steel bar is 2 7/8 inches thick. If 1/16 inch is milled from the bar, what is the new thickness?

21. 15 7/9 − 9 7/11 =

2. A loaded truck was found to weigh 8,472 1/4 pounds. The truck when empty weighed 3,549 3/4 pounds. What was the weight of the load?

22. 1005 3/8 − 799 4/5 =

23. 2 1/32 − 1 1/64 =

3. Find A.

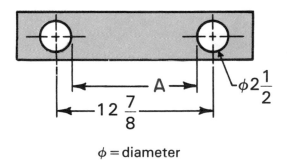

24. 5,275 − 214 5/9 =

ϕ = diameter

25. 12 9/17 − 5 2/34 =

4. No. 80 gauge wire is 27/1000 inch in diameter. No. 57 gauge wire is 43/1000 inch in diameter. How much larger in diameter is the No. 57 gauge wire than the No. 80 gauge wire?

8. Find A.

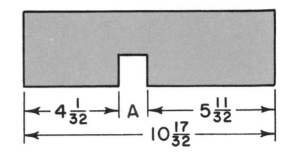

5. A carpenter had a board 34 3/4 inches long. To fit the space for a shelf, 7/16 inch was cut off one end. How long was the board after the piece was removed?

9. Find A.

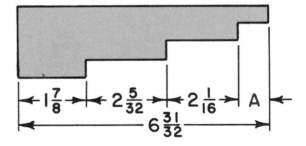

6. The outside diameter of a pipe is 1 7/8 inches, the wall of the pipe is 3/16 inch. Find the inside diameter.

7. A tapered shaft is 2 7/16 inches in diameter at one end and 1 3/32 inches at the other. What is the difference in diameter?

10. An electrode in an arc welder was 8 11/16 inches in length. After use, the electrode measured 3 3/8 inches. What was the length of the portion used?

MULTIPLICATION OF FRACTIONS, MIXED, WHOLE NUMBERS

To multiply two fractions, first multiply the two numerators to get the numerator for the answer, then multiply the two denominators to get the denominator for the answer. Write the product (answer) of the numerators over the product of the denominators, and reduce the fraction to its lowest form.

Problem: Multiply 3/4 × 2/3.

Solution:

$$\frac{3 \text{ times } 2}{4 \text{ times } 3} = \frac{6}{12} = \frac{1}{2}$$

To multiply a whole number times a fraction, first change the whole number to a fraction and multiply as two fractions. For example, the whole number 2 is the same as $\frac{2}{1}$.

Problem: Multiply $3 \times \frac{5}{16}$.

Solution:

$$3 \text{ or } \frac{3 \text{ times } 5}{1 \text{ times } 16} = \frac{15}{16}$$

EXERCISE 10-1

Multiply the following. Reduce to lowest terms. Show your work.

1. $\frac{1}{2} \times \frac{4}{5} =$

2. $\frac{2}{3} \times \frac{3}{8} =$

3. $\frac{5}{9} \times \frac{3}{10} =$

4. $\frac{2}{3} \times \frac{5}{11} =$

5. $\frac{3}{4} \times \frac{3}{7} =$

6. $\frac{1}{8} \times \frac{3}{16} =$

7. $\frac{24}{5} \times \frac{1}{4} =$

8. $\frac{3}{4} \times \frac{7}{8} =$

9. $\frac{9}{10} \times \frac{3}{5} =$

10. $\dfrac{5}{9} \times \dfrac{1}{8} =$

18. $\dfrac{35}{4} \times \dfrac{4}{35} =$

11. $\dfrac{1}{6} \times \dfrac{7}{12} =$

19. $\dfrac{2}{3} \times 0 =$

12. $\dfrac{15}{32} \times \dfrac{1}{8} =$

20. $\dfrac{7}{3} \times 8 =$

13. $9 \times \dfrac{1}{9} =$

21. $\dfrac{9}{5} \times \dfrac{2}{3} =$

14. $\dfrac{1}{26} \times 26 =$

22. $\dfrac{33}{14} \times \dfrac{7}{99} =$

15. $92 \times \dfrac{1}{51} =$

23. $\dfrac{24}{13} \times \dfrac{26}{12} =$

16. $\dfrac{7}{4} \times \dfrac{1}{7} \times 7 =$

24. $\dfrac{77}{15} \times 5 =$

17. $\dfrac{4}{5} \times 3 =$

25. $\dfrac{53}{64} \times \dfrac{9}{32} =$

EXERCISE 10-2

Solve these problems. Show your work.

1. A milling machine is set to remove 1/64 inch of metal with each pass. What will be the total thickness of metal removed with 12 successive cuts?

2. If the diameter AB of the circle is 15/16 inch and the radius OC is half that long, find the distance of the radius OC.

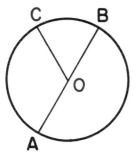

3. Thirteen strips, 3/4 inch wide, are to be ripped from a sheet of plywood. If 1/8 inch is lost with each cut, how much of the plywood sheet is used in making the 13 strips? (Assume 13 cuts are necessary.)

4. A steel rod is 15 inches long. A piece of the rod equal to 1/20 of its total length is cut off. How long is this piece?

5. Sheets of copper to be used as roof flashing are 1/16 inch thick. If 24 of these copper sheets are stacked, what is the thickness of the stack?

6. Six inches of wire solder weighs 9/16 oz. What is the weight of one inch of wire solder?

7. How much stock will be needed to cut 28 pieces, 3/4 inch long, allowing 1/16 inch for each saw cut? (Assume 28 cuts are necessary.)

8. Thirty-six pins, 5/8 inch long, are to be cut from a bar of metal. Allowing 1/16 inch waste per pin, how much bar stock is required? (Assume 36 cuts are necessary.)

9. The plate illustrated is to be drilled with five holes equally spaced. Find the length of the plate.

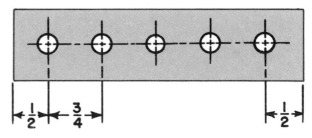

MULTIPLICATION—FRACTION BY MIXED, WHOLE NUMBER

To multiply a fraction by a mixed number, change the mixed number to an improper fraction (numerator is larger than denominator). Multiply the numerator of the improper fraction by the numerator of the common fraction, then multiply the two denominators. Reduce to lowest form.

Problem: Multiply $\dfrac{1}{2} \times 4\dfrac{3}{4}$

Solution: $\dfrac{1}{2} \times \dfrac{19}{4} = \dfrac{19}{8} = 2\dfrac{3}{8}$

To multiply two mixed numbers, change both to improper fractions, and proceed as described in the preceding paragraph.

EXERCISE 10-3

Multiply the following. Reduce to lowest terms. Show your work.

1. $3/4 \times 6\,1/2 =$

2. $5/8 \times 36 =$

3. $6/7 \times 5\,4/9 =$

4. $4\,16/75 \times 10\,5/32 =$

5. $9\,3/7 \times 2\,1/2 =$

6. $18 \times 4/9 =$

7. $17\,3/5 \times 2\,5/8 =$

8. $4/7 \times 1\,3/4 =$

9. $16/25 \times 1\,4/5 =$

10. 15 2/3 × 8 5/16 =

16. 3 1/3 × 3 2/7 =

11. 3 10/21 × 5 7/16 =

17. 8 1/3 × 6 3/4 =

12. 5 10/27 × 9 3/8 =

18. 2/3 × 2 3/4 =

13. 6 3/4 × 8 3/4 =

19. 1/5 × 3 1/2 =

14. 4 5/18 × 3 3/14 =

20. 7 1/3 × 1 1/2 =

15. 3 1/2 × 7 1/3 =

21. 2 5/16 × 5 5/12 =

EXERCISE 10-4

Solve these problems. Show your work.

1. A salesperson worked on a job for 10 weeks, 5 1/2 days per week, and 9 1/4 hours per day. How many hours did the salesperson work during this time period?

2. If 1 3/8 inch lengths of steel wire are used in making nails, how many inches of wire will be required to make 1,000 nails?

3. A cosmetologist, who works in a neighboring town, drives 7 1/2 miles one way to her place of employment. If she works five days per week, what is her weekly mileage to and from work?

4. A special type washer is stamped from a steel strip. If 15/16 inch is required for one washer, how much of the steel strip will be required to make 5,000 washers?

5. The area of a rectangle is found by multiplying the length of the rectangle times its width. What is the area of a rectangle whose length is 3 7/8 inches and width is 2 5/16 inches?

6. Twenty-five pieces of wire, each 27 3/4 inches long, are cut from a roll of wire. What is the total number of inches of wire used?

7. A pump raises 247 3/4 gallons of water per hour. How much water will it raise in 6 1/2 hours?

8. A power shovel can move 1 1/3 cubic yards of earth per minute. How much earth will it move in 5 1/2 minutes?

9. Two trucks are used in hauling crushed rock. The larger truck hauls 3 1/2 cubic yards per load and the other hauls 2 1/4 cubic yards per load. If each truck hauls 16 loads, what is the total quantity of crushed rock hauled in cubic yards?

10. A stationary gasoline engine requires a special gasket that is cut from a roll of material. Each gasket requires 11 3/4 inches. What length of material is required to make gaskets for 4 engines?

11. A strip of 1 × 2 lumber is cut into 10 5/16 inch lengths for shelf cleats. If 1/8 inch is lost with each cut, how much length is used in making 12 cleats? (Assume that 12 cuts are necessary.)

12. There are 15 risers in a set of stairs running from the basement to the first floor of a residence. What is the distance between floors?

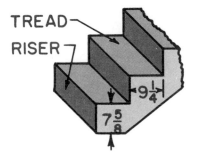

TREAD
RISER

13. There are 14 treads in the same set of stairs. What is the total width of the treads?

14. The net weight of a box of machine screws and nuts is 2 3/8 pounds. A hardware dealer has 24 1/2 boxes in stock. What is the total weight of these screws and nuts?

15. An irrigation ditch has 30 holes that need to be plugged. If each hole requires 2/3 of a cubic foot of concrete to plug it, what is the total amount of concrete required?

16. An area next to a building must be fenced. It will require 3 sides to be fenced, each of which is 17 1/2 meters long. What is the total length of fencing required?

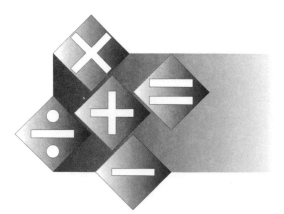

DIVISION OF FRACTIONS, MIXED, WHOLE NUMBERS

In the division of fractions, we invert the divisor (turn divisor upside down) then proceed as in multiplication of fractions.

Problem: $\dfrac{3}{8} \div \dfrac{3}{4}$

Solution:

$$\frac{3}{8} \div \frac{3}{4} = \frac{3}{8} \times \frac{4}{3} = \frac{12}{24} \text{ or } \frac{1}{2}$$

Mixed numbers are first changed to improper fractions before inverting the divisor and multiplying.

Problem: $2\dfrac{1}{2} \div \dfrac{2}{3}$

Solution:

$$2\frac{1}{2} \div \frac{2}{3} = \frac{5}{2} \times \frac{3}{2} = \frac{15}{4} \text{ or } 3\frac{3}{4}$$

If the divisor is a whole number, we invert the divisor by writing 1 over it, and multiplying.

Problem: $\dfrac{3}{4} \div 5$

Solution:

$$\frac{3}{4} \div 5 = \frac{3}{4} \div \frac{5}{1} = \frac{3}{4} \times \frac{1}{5} = \frac{3}{20}$$

EXERCISE 11-1

Divide the following. Reduce to lowest terms. Show your work.

1. 5/8 ÷ 3/6 =

2. 6/9 ÷ 2/3 =

3. 3/4 ÷ 51/16 =

4. 36 ÷ 7 1/5 =

5. 25 ÷ 3 1/3 =

6. 18 ÷ 1/8 =

7. 15 ÷ 7/12 =

8. 33 ÷ 22/7 =

9. 3/4 ÷ 1/8 =

10. 14/3 ÷ 7/4 =

11. 4 1/3 ÷ 3/8 =

12. 6 1/4 ÷ 7/10 =

13. 8 1/2 ÷ 3/4 =

14. 36/25 ÷ 18/5 =

15. 1/3 ÷ 1/3 =

16. 9/10 ÷ 7 =

17. 1 1/2 ÷ 4 =

18. 5/6 ÷ 2 =

19. 6 2/3 ÷ 2 =

20. 6 1/7 ÷ 4 =

21. 2 1/4 ÷ 1 4/5 =

22. 12 1/2 ÷ 4 1/4 =

23. 1 7/9 ÷ 5 1/4 =

24. 50 2/3 ÷ 25 1/3 =

25. 3 1/3 ÷ 5/9 =

26. 4 1/6 ÷ 6 1/3 =

27. 5 1/4 ÷ 2 3/3 =

28. 5 3/8 ÷ 3 1/4 =

29. 6 31/32 ÷ 4 =

30. 9 33/64 ÷ 7/8 =

EXERCISE 11-2

Solve these problems. Show your work.

1. How many pieces of metal, each 3/4 inch in length can be cut from a strip 75 inches long? Assume there is no waste in cutting.

2. A worker can produce one welded bracket in 2/3 hour. How many of these brackets can be produced in 8 hours?

3. A truck is loaded with cartons each weighing 1/20 ton. If the total weight of the cartons is 1 3/4 tons, how many cartons are on the truck?

4. How many pieces of wire, each 2/3 foot long, can be cut from a roll of wire 15 feet in length? How much wire remains after the last piece is cut?

5. A roll of cloth, 42″ wide and 200′ long, is being considered for use in production of tablecloths that require 44″ of material each. How many tablecloths can be produced? How much cloth would be left over?

6. How many shelves, each 3 1/2 feet long, can be cut from a board 14 feet long? How much is left?

7. A drill will bore through 35 inches of steel before it needs to be sharpened. How many holes can be drilled through a 7/8 inch steel plate with this drill before resharpening?

8. If 1/9 cubic yard of cement is needed in mixing one cubic yard of concrete, how many cubic yards of concrete can be mixed with 7/8 cubic yard of cement.

9. The height of a wall raises 5 1/4 inches with each course of bricks including the motar joint. How many courses of bricks would be needed to build a wall 105 inches high?

10. In turning a shaft on a metal lathe, the cutting tool moves 1/2 inch per minute. How much time will it take the tool to travel the length of an 18 3/4 inch shaft?

11. How many sheets of metal, each 1/32 inch thick, are there in a pile 8 3/8 inches high?

12. How many lengths of radiator hose, each 3 3/16 inches long, can be cut from a 38 1/4 inch length of hose? Disregard waste in cutting.

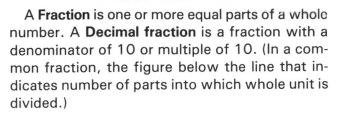

WRITING AND READING DECIMAL FRACTIONS

A **Fraction** is one or more equal parts of a whole number. A **Decimal fraction** is a fraction with a denominator of 10 or multiple of 10. (In a common fraction, the figure below the line that indicates number of parts into which whole unit is divided.)

In writing decimal fractions, we omit the denominator and place a dot, called a **Decimal point,** in front of the numerator (number of parts of unit with which we are concerned). The denominator of a common fraction may be any number. The denominator of a decimal fraction, which is unwritten, is always a multiple of 10, such as 10, 100, 1,000, 10,000, etc.

For many purposes, decimal fractions (decimals) are easier to write and to compute than common fractions. Examples of decimal usage: Automobiles register the distance traveled in decimals, gasoline pumps measure in tenths, precision measurements (micrometer), and our monetary system.

HOW TO WRITE AND READ DECIMALS

Fractions such as 2/10 and .2 are both read the same, two tenths. When there is one figure at the right of the decimal point, we read the decimal as tenths. If there are two figures at the right of the decimal point, the decimal is in hundredths; three figures thousandths; four figures ten thousandths.

Fraction	Decimal	
$\dfrac{2}{10}$	.2	Two tenths
$\dfrac{2}{100}$	.02	Two hundredths
$\dfrac{2}{1,000}$	.002	Two thousandths
$\dfrac{2}{10,000}$	.0002	Two ten thousandths

In writing decimals, such as .2, the value of .2 (two tenths) is ten times as great as it would have been if it had been written one more place to the right, .02 (two hundredths). Zeros are used to fill the empty space. . .as place holders.

If the number has figures both to the left of the decimal point, and to the right, the number is called a **Decimal mixed number.** The decimal mixed number 4.165 is read four and one hundred sixty five thousandths. The word "and" is used to separate the whole number from the decimal fraction.

In lab or field work, the decimal point is often spoken of as point. The number 4.165 is read four point one six five.

EXERCISE 12-1

Write the following common fractions as decimals, and decimal mixed numbers:

1. 3/10 _____.

2. 75/1,000 _____.

3. 31/10,000 _____.

4. 46/100 _____.

5. 888/1,000 _____.

6. 2 4/10 _____.

7. 117 33/1,000 _____.

8. 72/100 _____.

9. 2,174 3/10 _____.

10. 93 17/10,000 _____.

11. 82/1,000 _____.

12. 34/100 _____.

13. 892 3/10 _____.

14. 97/1,000 _____.

15. 221/1,000 _____.

16. 8/10,000 _____.

17. 2 9/10 _____.

18. 163 91/1,000 _____.

19. 27/100 _____.

20. 911 3/10 _____.

EXERCISE 12-2

Write the following in decimals, and decimal mixed numbers:

1. Two tenths _____.

2. Twenty-four hundredths _____.

3. Three hundred forty-five thousandths

_____.

4. Thirty-five ten thousandths _____.

5. Twenty-four thousandths _____.

6. Three hundred eighty-seven thousandths

_____.

7. Two hundred fifty hundred thousandths

_____.

8. Two hundred three thousandths_____.

9. Ninety-nine thousandths _____.

10. Nine thousandths _____.

11. One thousand twenty-five ten thousandths

_____.

12. Two thousand fifty hundred thousandths

_____.

13. Five and fourteen ten thousandths

_____.

14. Twenty-four and six hundred three thousandths _____.

15. Fourteen and two hundredths _____.

16. Five and three thousand one hundred twelve ten thousandths _____.

17. Sixty-five thousandths _____.

18. Twenty-three and six ten thousandths

_____.

19. Seventy-five ten thousandths _____.

20. Four hundred seventy-five thousandths

_____.

21. Six hundred sixty-six ten thousandths

_____.

22. Fifty-four thousandths _____.

23. Two hundred sixty-four ten thousandths

_____.

24. Seventy-nine and three tenths_____.

25. Forty-one and sixty-three thousandths

_____.

26. Two hundred sixty-eight ten thousandths

_____.

27. Thirty-nine and two hundredths _____.

28. One thousand one and one ten thousandths

_____.

29. Twenty-two thousandths _____.

30. Ninety-seven ten thousandths_____.

WRITING AND READING DECIMAL FRACTIONS

EXERCISE 12-3

Change the following to decimal fractions or mixed decimals:

1. 3 4/10 _____.

2. 27/100 _____.

3. 349/1,000 _____.

4. 117 711/10,000 _____.

5. 43 3/1,000 _____.

6. 1,016 3/10 _____.

7. 3/10,000 _____.

8. 99/1,000 _____.

9. 9/1,000 _____.

10. 1,025/10,000 _____.

11. 12/10,000 _____.

12. 2,050/10,000 _____.

13. 60 14/100 _____.

14. 14/10,000 _____.

15. 603/10,000 _____.

16. 96/1,000 _____.

17. 93 6/100 _____.

18. 172/10,000 _____.

19. 26/1,000 _____.

20. 119 91/100 _____.

21. 475/1,000 _____.

22. 23 9/10 _____.

23. 66/100 _____.

24. 6/10,000 _____.

EXERCISE 12-4

Change the following as indicated. Reduce to simplest terms.

1. A dowel is .5 inch in diameter. Express this as a fraction.

2. A piece of plywood is 75/100 inch thick. Express this as a decimal fraction.

3. A strip of aluminum is cut so it is .907 inch wide. Express this as a fraction.

4. A setscrew is .375 inch long. Express this as a fraction.

5. A wood screw weighs 242/1,000 ounce. Express this as a decimal fraction.

6. The thickness of a template is .145 inch. Express this as a fraction.

7. A metal disk is 61/1,000 inch thick. Express this as a decimal fraction.

8. An insulated wire has a diameter of .091 inch. Express this as a fraction.

9. A vertical milling machine removes .0112 inch of metal with each stroke. Express this as a fraction.

10. An appliance repair person used .948 ounce of solder on a job. Express this as a fraction.

11. A strip of sheet metal weighs 5 214/1,000 pounds. Express this as a mixed decimal.

12. An abrasive machine removed 23/10,000 inch from one side of a steel plate. Express this as a decimal.

13. A transistor has a thickness of .0305 inch. Express this as a fraction.

14. A casting weighs 2.993 pounds. Express this as a mixed numeral.

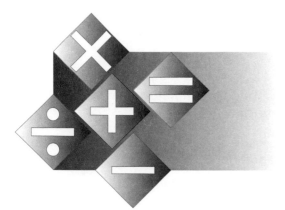

ADDITION OF DECIMALS

In adding decimals, the only factor that is different from adding whole numbers, is the placing of the decimal point. We write the figures to be added so the decimal points are in a straight vertical column (one is placed directly under the other). We add and carry from one column to the other as in adding whole numbers and place a decimal point in the sum, directly beneath the decimal points above.

Examples:

```
  .26          4.607
  .168          .408
  .428         5.015
```

EXERCISE 13-1

Add the following problems:

1.
```
   .024
   .165
```

2.
```
   .873
   .199
```

3.
```
   4.6
   4.62
```

4.
```
   .0003
   .0099
```

5.
```
   5.00
    .99
```

6.
```
   4.5
    .008
```

7.
```
   62.14
    8.87
```

8.
```
   3.487
   9.645
```

9.
```
   126.7
    81.09
```

10.
```
   1.544
   3.609
```

11.
```
   .4999
   .5664
```

12.
```
   80.95
   36.26
```

13.   .987
       .765

14.   1.7
       2.4
       3.0

15.   5.8
       .9
       6.3

16.   14.2
       13.6
        8.4

17.   21.5
       18.9
       32.8

18.   .0131
       .1640
       .2229

19.   8.69
       3.07
       17.50

20.   4.612
       .958
       .885

21.   5.29
       4.38
       9.62

22.   72.24
       16.38
       92.37

23.   808.01
       146.92

24.   700.08
       145.67

25.   600.0009
        50.9965

26.   4187.75
       1834.85

27.   168.8301
        23.0009

28. $23.97
       14.34

29. $108.33
     124.67

30. $1986.45
     312.45
     1000.98

31. 78.0035
     22.7809
     44.3765
     109.0023

32. 1005.7
     230.076
     59.2308
     110.0

33. $4987.23
     4980.47
     10234.97
     9872.54

34. $500.69
     42.33
     103.02
     49.95

35. .52802
     .34007
     .00321
     .00009

36. 15.125
     4.625
     20.125
     16.125

EXERCISE 13-2

Solve the following problems. Show your work.

1. An electrician cut the following lengths of wire from a roll: 3.7, 2.65, and 6.8 meters. What was the total amount of wire cut from the roll?

2. A maintenance foreman wishes to fence a rectangular area for storage. If the area is 26.5 feet wide and 44.25 feet long, how much fence will be needed?

3. The thicknesses of several pieces of metal were measured with a micrometer and found to be: .035, .608, .0513, and .026 inch. What is the combined thickness of the four pieces of metal?

4. A television repairer earned the following amounts in five days: $296.65, $306.44, $298.72, $285.38, $304.29. What was the total earned for the week?

5. A truck driver loaded three cartons on the truck. The weights of the cartons were: 33.45, 43.5, and 242.4 kilograms. What was the total weight of the three cartons?

6. A rod is to be made of tool steel 5.750 inches in length. It will be acceptable if it varies from this length by no more than .003 inch. What is the longest rod that will be acceptable?

7. A plumber worked for one week on a construction job. The number of hours worked each day were: 10.75, 8.5, 9.33, 8.0, and 6.25. How many hours were worked that week?

8. A cabinetmaker built a set of bookshelves. (See diagram.) What was the total height of the shelves in inches?

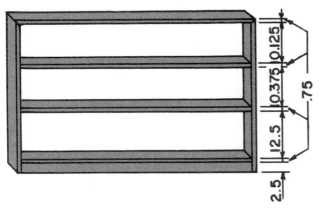

9. A plumber cut the following lengths of 3/4 inch pipe: 12.375 inches, 8.625 inches, 48.5 inches, 33.875 inches. What was the total length of pipe used? (Disregard waste in cut.)

10. A surveyor, in measuring an irregular plot of land, recorded the following measurements: 175.33 feet, 201.875 feet, 89.67 feet, 145.125 feet, and 110.375 feet. What was the total distance around the plot of land?

The **Perimeter** of a geometric plane figure is found by adding the lengths of its sides. Find the perimeters of the figures in problems 11 through 16. The dimensions given are in centimeters.

11.

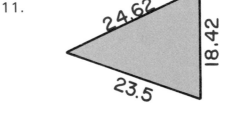

12.

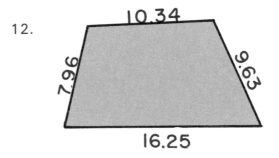

13.

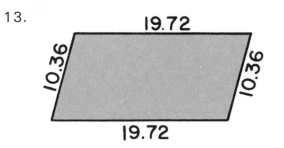

19.72

10.36

10.36

19.72

15.

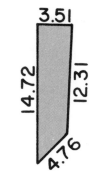

3.51

14.72

12.31

4.76

14.

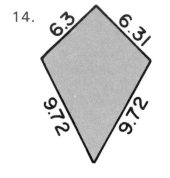

6.3

6.31

9.72

9.72

16.

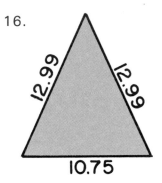

12.99

12.99

10.75

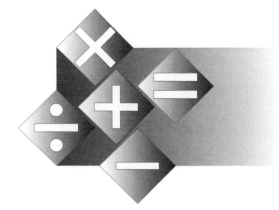

SUBTRACTION OF DECIMALS

To subtract decimal fractions, write the decimals in vertical form and subtract them as if they were whole numbers. We must be careful to see that the numerals are written with their decimal points in corresponding positions. The decimal point in the difference is placed directly below the decimal points in the subtrahend and the minuend.

Examples:

```
  .658        .923        2.343
  .234        .869         .876
  .424        .054        1.467
```

EXERCISE 14-1

Subtract the following problems:

1.   .74
    .43

2.   .84
    .34

3.   4.6
    .9

4.   23.9
    12.8

5.   5.0
    3.5

6.   8.1
    2.2

7.   7.4
    2.5

8.   8.0
    2.3

9.   9.2
    2.8

10.   7.5
    1.8

11.   4.13
    1.91

12.   6.23
    2.32

13.   7.42
    2.89

14.   8.13
    2.51

15.	7.49 2.33	25.	17.648 8.096	35.	.609375 .359375
16.	8.50 2.47	26.	31.4802 7.0165	36.	.765625 .25
17.	8.24 3.25	27.	9.60410 3.71521	37.	8.3214 .0137
18.	9.32 2.45	28.	11.68102 11.33092	38.	42.0666 2.3845
19.	8.17 1.29	29.	30.001 29.776	39.	3.21167 .61582
20.	7.205 3.419	30.	867.514 66.904	40.	12.4992 .8078
21.	6.801 3.743	31.	421.868 227.872	41.	.8995 .0228
22.	32.484 9.585	32.	506.432 416.846	42.	.00972 .00454
23.	56.873 48.004	33.	.515625 .484375	43.	7.38 .063
24.	87.9070 58.6707	34.	.265625 .140625	44.	9.0324 7.1

45. $16.78 - 5.99 =$

46. $.9453 - .765 =$

47. $2.345 - 1.6 =$

48. $.13 - .023 =$

49. $.00754 - .00093 =$

50. $20.32 - .84 =$

51. $930. - 17.45 =$

52. $.034 - .034 =$

53. $1. - .9999 =$

54. $7.777 - .777 =$

EXERCISE 14-2

Solve these problems. Show your work.

1. Two sheets of metal were measured with a micrometer. One was found to be .093 inch thick and the other was .133 inch thick. What is the difference in the thickness of the two sheets of metal?

2. A plumber cut 7.25 centimeters off the end of a piece of pipe. The piece of pipe was 86.125 centimeters long. What is the length of the remaining pipe?

3. A cut of .035 inch was milled from a metal block 3.875 inches in thickness. What is the new thickness of the block?

4. A carpenter's bill for lumber totaled $125.67. The yard gave a discount of $12.57. How much did the carpenter pay for the lumber?

5. A machine is used to cut circular disks from sheet metal. If 265 kilograms of sheet metal is necessary to make 206.25 kilograms of disks, how much of the metal is waste?

6. A truck driver loaded 10,000 pounds of sand on the truck. When the load was weighed in at the destination, there was only 9,874.75 pounds of sand. How much sand was lost on the trip?

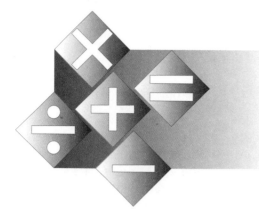

MULTIPLICATION OF DECIMALS

In multiplying numbers involving decimals, write the numbers under each other the same as in the multiplication of whole numbers. The decimal points are included but are disregarded until the multiplication has been completed.

Examples:

```
1.    .23
      .6
     .138
```

To solve problem 1, we multiply as if the numbers were whole numbers 23 and 6. Then we count the total number of places to the right of the decimal points in .23 and .6. There are two places to the right of the decimal point in .23, and one place to the right of the decimal point in .6. This is the total of three places. We count three places to the left starting with number 8 in 138 and place the decimal point, giving us .138 for our answer.

```
2.    .412
      .63
     1236
     2472
    .25956
```

We have multiplied as if the numbers were 412 and 63. Then we count the total number of places to the right of the decimal points in .412 and .63, giving us a total of five. We count five places to the left starting with the 6 in 25956 and place the decimal point. The answer is .25956.

```
3.    4.13
      1.03
     1239
     000
     413
    4.2539
```

In problem 3, we multiply mixed decimal numbers. We multiply as if the numerals were 413 and 103. In counting the decimal places to the right of the decimal points we find there are four. We count four places to the left starting with number 9 in the product 42539, giving us 4.2539 for our answer.

EXERCISE 15-1

Multiply the following decimal fractions. Show your work.

```
1.    .68
       6
```

```
2.    6.87
       .21
```

```
3.    .39
       5
```

```
4.    .11
      .35
```

MULTIPLICATION OF DECIMALS

5.
```
   8.52
      4
```

6.
```
   7.02
      6
```

7.
```
   .57
   .4
```

8.
```
   .97
   .2
```

9.
```
   14.18
     .64
```

10.
```
   69.5
    .08
```

11.
```
   39.2
    .29
```

12.
```
   8.44
    .04
```

13. 48.433 × 10 =

14. .235 × 100 =

15. 38.86 × 1,000 =

16. 25.9 × 10,000 =

17. 7.53 × 7.98 =

18. .836 × .9 =

19. 1.594 × 21.47 =

20. 700.4 × .16 =

21. .0087 × 23 =

Solve these problems. Show your work.

1. A truck driver loaded 4 cartons each weighing 29.5 kilograms, and 5 cartons each weighing 48.25 kilograms on the truck. What was the total weight of the 9 cartons?

22. 999.1 × 11.9 =

2. The area of a rectangle is found by multiplying its length times its width. What is the area of a rectangular piece of sheet metal 16.4 inches long and 12.16 inches wide?

23. 13.65 × .0001 =

3. A carpenter used 7.75 sheets of interior plywood on a job. If the plywood cost $23.40 per sheet, what was the total cost of the plywood used on the job?

24. .263 × .0032 =

4. A rectangular storage tank has an inside length of 2.75 meters and an inside width of 2.375 meters. What is the total number of square meters (m²) of area in the base of the tank?

25. 23.4 × .9001 =

5. Fifty-two sheets of metal are stacked one on top of the other. If each sheet is .085 inch thick, what is the total height of the sheets?

26. 1.003 × 2.005 =

6. At $1.84 per foot, what is the cost of 24.5 feet of steel tubing?

27. .233 × 9 =

7. If a 7/8 inch rivet weighs .189 pounds, what is the weight of 750 such rivets?

DIVISION OF DECIMALS

In dividing with decimals we proceed as in the division of whole numbers, then we see to it that the decimal point is properly placed.

DIVIDING DECIMAL BY WHOLE NUMBER

Problem: Divide .96 by 4

```
      .24
  4/.96
     8
    16
```

When a decimal is divided by a whole number, the number of decimal places in the quotient is the same as the number of decimal places in the dividend. The decimal point in the quotient is placed directly above the decimal point in the dividend.

DIVIDING WHOLE NUMBER BY DECIMAL (No Remainder)

Problem: Divide 96 by .4

```
        240.
  .4ʌ/96.0ʌ
      8
     16
     16
```

In dividing a whole number by a decimal, we annex (add to) the dividend as many zeros as there are decimal places in the divisor. The quotient is then a whole number. Use the symbol "ʌ" to mark the new position of the decimal point in the dividend and the divisor.

DIVIDING WHOLE NUMBER BY DECIMAL (With Remainder)

Problem: Divide 99 by .4

```
        247.5
  .4ʌ/99.00ʌ
      8
     19
     16
      3 0
      2 8
        20
        20
```

In dividing a whole number by a decimal, if there is a remainder we add a decimal point, then annex zeros and continue the division. The quotient or answer is then a decimal mixed number. The number of decimal places in the dividend minus the number of decimal places in the divisor indicates the number of places in the quotient.

DIVIDING A DECIMAL BY A DECIMAL

Problem: Divide 9.968 by 1.4

```
          7.12
  1.4ʌ/9.9ʌ68
       9 8
        16
        14
         28
         28
```

In this problem, the number of decimal places in the dividend (three) minus the number of decimal places in the divisor (one) equals the number in the quotient or answer (two).

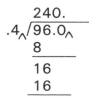

CHECKING DIVISION PROBLEM ANSWERS

5. .6 ÷ 3 =

Problems in the division of decimals may be checked by multiplying the quotient by the divisor and comparing the result with the dividend.

Example:

```
              Check
   240.        240
.4/ 96.0ᐱ       .4
    8          96.0
   16
   16
```

6. 1.4 ÷ .2 =

EXERCISE 16-1

7. 3.2 ÷ .08 =

Divide the following problems. Show your work.

1. .8 ÷ .2 =

8. .12 ÷ .04 =

2. .9 ÷ .3 =

9. .28 ÷ .07 =

3. .6 ÷ .3 =

4. .8 ÷ 4 =

10. 25.84 ÷ 19 =

11. $2576 \div 1.12 =$

17. $8487 \div .23 =$

12. $.03456 \div .04 =$

18. $84.87 \div 2.3 =$

13. $1 \div 5 =$

19. $6.111 \div .97 =$

14. $2 \div 4 =$

20. $6975 \div .93 =$

15. $12 \div .06 =$

21. $.7832 \div .08 =$

16. $.8487 \div .23 =$

22. $.0288 \div .012 =$

23. 15.18 ÷ 2.3 =

Solve these problems. Show your work.

1. A machinist worked 8.5 hours and earned $182.75. What is the machinist's hourly rate?

24. .000072 ÷ .24 =

2. A carpenter sawed a board 15.75 feet long into equal pieces, each 1.75 feet long. Disregarding the waste in cutting, how many pieces were made?

25. 2.0878 ÷ 26 =

3. A computer manufacturing clerk, responsible for packing and sending computer parts, was told there were 1024 parts to be packed. The parts were to be packed 32 per carton. How many cartons were needed?

26. 73.275 ÷ 2.5 =

4. A builder divided 21.6 acres of land into lots each .8 acres in area. How many lots did this provide?

27. .0057 ÷ 19 =

5. One cubic yard of concrete will pour 81 square feet of concrete slab of the desired thickness. How many cubic yards of concrete will be needed to pour a slab of the same thickness containing 502.2 square feet?

9. A fence is constructed with boards 9.75 inches wide. If the boards are placed vertically, with no space between, how many are required to build the fence 71.5 feet long?

6. A contractor removed 35.7 cubic meters (m³) of earth from a building site. If the trucks can haul 1.7 m³ per load, how many truck loads of earth were removed?

10. A pump will remove 29.75 liters (ℓ) of water per minute from a basement excavation. How many minutes would be required to remove 1788.57 liters?

7. A machine produced 495 cotter pins in 8.25 minutes. What is the average number of cotter pins that it can produce per minute?

11. A contractor has a fleet of five trucks. In a five day week the trucks traveled a total of 4887.5 kilometers. What was the average distance traveled by each truck per day?

8. A strip of metal 165 inches long is to be divided into 21 equal pieces. What will be the length of each piece? (Carry answer to 3 places.)

12. One thousand machine bolts weigh a total of 586 pounds. What is the weight of each bolt?

CHANGING COMMON FRACTIONS TO DECIMALS

To change a common fraction to a decimal, annex one or more zeros to the numerator, and divide by the denominator. If the number does not come out even, it may be carried as many decimal places in the answer as desired.

Problem: Change 7/8 to a decimal fraction.

```
      .875
8 / 7.000
    6 4
    ────
      60
      56
    ────
      40
      40
    ────
```

When changing a common fraction to a decimal fraction, the number of decimal places in the quotient must be the same as the number of zeros that were annexed to the numerator (dividend).

Problem: Change 2/7 to a decimal fraction.

```
      .2857
7 / 2.0000
    1 4
    ────
      60
      56
    ────
      40
      35
    ────
      50
      49
    ────
       1
```

When the division does not come out even, we carry it far enough to get as many decimal places as may be desired in the answer.

To change mixed numerals to mixed decimals, first change to an improper fraction and divide as with common fractions.

Problem:

$$3\frac{3}{4} = \frac{15}{4} = $$

```
         3.75
4 / 15.00
    12
    ────
     30
     28
    ────
      20
      20
    ────
```

EXERCISE 17-1

Change these common fractions to decimal fractions. For problems that do not come out even, carry to the nearest thousandth, unless otherwise indicated. Show your work.

1. $\dfrac{1}{64}$

2. $\dfrac{1}{32}$

3. $\dfrac{3}{64}$

8. $\dfrac{5}{32}$

4. $\dfrac{1}{16}$

9. $\dfrac{13}{64}$

5. $\dfrac{5}{64}$

10. $\dfrac{1}{4}$

6. $\dfrac{3}{32}$

11. $\dfrac{5}{16}$

7. $\dfrac{1}{8}$

12. $\dfrac{11}{32}$

13. $\dfrac{3}{8}$

14. $\dfrac{25}{64}$

15. $\dfrac{15}{32}$

16. $\dfrac{31}{64}$

17. $\dfrac{1}{2}$

18. $\dfrac{37}{64}$

19. $\dfrac{3}{4}$

20. $\dfrac{51}{64}$

21. $\dfrac{7}{8}$

22. $\dfrac{59}{64}$

EXERCISE 17-2

Changing fractions to decimal fractions. Solve these problems. Show your work.

1. A wall consists of cement blocks 7 5/8 inches thick and plaster (one side) 3/4 inch thick. Write the total thickness of the wall as a mixed decimal.

2. A salesperson sold four pieces of upholstery binding tape in the following lengths: 50", 42 1/2", 37 3/8", and 27 5/8". Write the total length of the tape sold as a mixed decimal.

3. A piece of plywood is made up of one piece of 1/4 inch wood, two pieces of 1/8 inch wood, and two pieces of 1/16 inch wood. Write the total thickness of the sheet of plywood as a decimal fraction.

4. A piece of hose has an inside diameter of 1 7/8 inches. The thickness of the wall is 5/32 inch. Express the outside diameter of the hose as a mixed decimal.

5. A machinist turned a 27/64 inch diameter shaft from a piece of stock 1/2 inch in diameter. Express the difference in diameter as a decimal.

6. A 10" wide board, used as a table leaf, is to be finished down to 7/8 of its original width. Express the new width as a mixed decimal.

7. A piece of steel 14 7/8 inches long was machined down to a finished length of 14 7/16 inches. Write the finished length as a mixed decimal.

9. There are 14 risers in a set of stairs. If each riser measures 8 1/8 inches, express the distance between the floors as a mixed decimal.

8. Insert the decimal equivalents for the common fractions in the drawing below. Round off the decimals to thousandths.

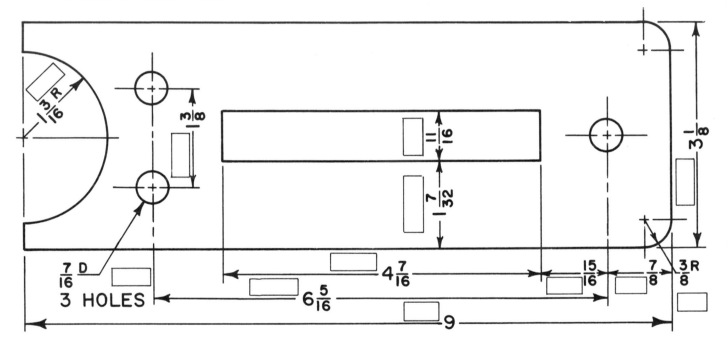

CHANGING DECIMALS TO COMMON FRACTIONS

In changing a decimal fraction to a common fraction, we drop the decimal point, write the denominator, then reduce the common fraction to its lowest form or terms.

Problem: Change .750 to a common fraction.

$$\frac{750}{1,000} \text{ or } \frac{3}{4}$$

For the denominator, use as many zeros as there are decimal places in the common fraction.

Problem: Change .21 to a common fraction.

$$\frac{21}{100}$$

This fraction is already in its lowest form, as it cannot be reduced.

EXERCISE 18-1

Change these decimals to common fractions. Show your work.

1. .375

2. .625

3. .75

4. .09375

5. .875

6. .50

11. .0625

7. .1875

12. .3125

8. .5625

13. .8125

9. .125

14. .6875

10. .15625

15. .9375

CONVERSIONS IN METRIC

As long as we are working in customary units (inch-pounds) and in metric units, it will be occasionally necessary to convert (to change) from one system of measurement to the other. You will also find it helpful to learn to convert from one unit in the metric system to multiples of that unit.

CONVERSION OF UNITS

The type of conversions most frequently occurring in customary and metric units are those in length, area, weight, volume, and temperature. These conversions can be accomplished by multiplying by a factor to get the answer or by checking for the answer in a conversion table. The multiplication factors are useful when tables are not available, but conversions from the tables are faster and reduce the chance of error.

LENGTH

To convert from customary units of length to metric units the procedure for multiplying is as follows: (Values approximate.)

To convert from	To	Multiply by
inches (in.)	millimeters (mm)	25.4*
inches (in.)	centimeters (cm)	2.54*
feet (ft.)	meters (m)	0.3048*
yards (yd.)	meters (m)	0.9144
miles (mi.)	kilometers (km)	1.6093
*exact conversions		

Example:

```
    2.625"
    25.4
   10500
   13125
   5250
   66.6750 mm
```

The procedure for converting customary and metric units from the tables in the Appendix, page 118 is as follows:

Using the same decimal as above (2.625"), find the value from the appropriate table for converting "Inches to Millimeters" and set the value down for 2 inches (50.8000). See below. Check the first column of the table and move down to .62, then to the right under the column .005 and set the value (15.8750) down below the other. Add the two conversion values. The conversion value is the same as found by the mathematical procedure.

```
2.625" =
2.000" = 50.8000 mm
0.625" = 15.8750 mm
         66.6750 mm
```

To convert from metric units to customary units, the procedure is as follows: (Values approximate.)

To convert from	To	Multiply by
millimeters	inches	0.0394
centimeters	inches	0.3937
meters	feet	3.2808
meters	yards	1.0936
kilometers	miles	0.6214

The procedures for multiplying or using the conversion tables for converting metric units to customary units are the same as shown earlier for converting customary units. Refer to the tables in the Appendix, page 117.

AREA

The procedure for converting customary and metric units of area are similar to those for length.

The conversion factors are as follows: (Values approximate.)

To convert from	To	Multiply by
square inches (in.²)	square centimeters (cm²)	6.5
square feet (ft.²)	square meters (m²)	0.09
square yards (yds.²)	square meters (m²)	0.84
square miles	square kilometers (km²)	2.6
acres	square hectometers (hm²)	0.4
square centimeters (cm²)	square inches (in.²)	0.16
square meters (m²)	square yards (yds.²)	1.2
square kilometers (km²)	square miles	0.4
square hectometers (hm²) (hectares)	acres	2.5

WEIGHT (MASS)

To convert customary and metric units of weight, the multiplication factors are as follows:

(Values approximate.)

To convert from	To	Multiply by
ounces (avoirdupois) (oz.)	grams (g)	28.3
pounds (lb.)	kilograms (kg)	0.45
tons (2000 lbs.)	megagrams (Mg)	0.9
grams (g)	ounces (oz.)	0.035
kilograms (kg)	pounds (lb.)	2.2
megagrams (Mg)	tons (2000 lbs.)	1.1

Conversion weight tables are included in the Appendix, page 122.

VOLUME

Customary and metric units of volume measure may be converted by the procedures of multiplication or the use of conversion tables.

Following are the multiplication factors for converting dry volume measure: (Values approximate.)

To convert from	To	Multiply by
cubic inches (in.³)	cubic meters (m³)	0.000 016
cubic feet (ft.³)	cubic meters (m³)	0.028 316
cubic yards (yd.³)	cubic meters (m³)	0.7646
cubic meters (m³)	cubic inches (in.³)	61 023
cubic meters (m³)	cubic feet (ft.³)	35.31
cubic meters (m³)	cubic yards (yd.³)	1.308

The conversion factors for liquid volume measure are as follows:

To convert from	To	Multiply by
ounces (apothecary) (oz.)	grams (g)	30.0
quarts (qt.)	liters (ℓ)	0.946
gallons (gal.)	liters (ℓ)	3.785
milliliters	ounces (oz.)	0.034
liters (ℓ)	quarts (qt.)	1.06
liters (ℓ)	gallons (gal.)	0.264

TEMPERATURE

The metric (SI) unit of temperature is the kelvin (K) but for practical applications Celsius (°C) is used. The degree intervals (100 degrees from ice point to boiling point of water) on the scale are the same for the kelvin scale and Celsius scale. The Fahrenheit scale has been used in the Customary System of measurement and has 180 degree intervals from ice point (32°) to boiling point (212°) of water.

Conversions in temperature readings in Fahrenheit and Celsius can be made by mathematics, as shown below. A table of common temperature conversions is shown on page 117 of the Appendix.

To convert from	To	Multiply by
degrees Fahrenheit	degrees Celsius	5/9 (after subtracting 32)
degrees Celsius	degrees Fahrenheit	9/5 (then add 32)

CONVERSION OF METRIC UNITS TO LARGER OR SMALLER METRIC UNITS

We learned in Unit 2 that the metric system is based on the decimal system and each unit is related to another by a multiple of 10. That is, a millimeter is $\frac{1}{10}$ of a centimeter, the centimeter is $\frac{1}{100}$ of a meter, and so on. When you know the values of the prefixes in the metric system, the conversion from one unit of metric to another is a matter of shifting the decimal point.

In converting from smaller units to larger units in metric, the decimal point is shifted to the left.

Example:

A millimeter is $\frac{1}{1000}$ of a meter

$$1000 \text{ millimeters} \div 1000 = \frac{1000}{1000} = 1 \text{ meter}$$

Example:

$$
\begin{array}{r}
2.875 \text{ m} \\
1000\overline{)2875.000} \text{ mm} \\
2000 \\
\hline
8750 \\
8000 \\
\hline
7500 \\
7000 \\
\hline
5000 \\
5000 \\
\hline
\end{array}
$$

What you really have done is move the decimal point three places to the left when dividing by 1000.

Example:

3016 mm = 3.016 m
65 mm = 6.5 cm (you divided by 10)

When converting from larger to smaller units in metric, the decimal point is shifted to the right.

Example:

1 meter equals 1000 millimeters

Also 3.617 meters
      1000

3 617,000 or 3617 millimeters

The same procedure of shifting the decimal point is used in other units of measure in the metric system. For example:

1268 grams   = 1.268 kilograms
  37 deciliters = 3.7 liters
  50 dekaliters = 500 liters

The approved metric prefixes and their decimal equivalents are shown on page 118 in the Appendix.

EXERCISE 19-1

Convert the following by use of their multiplication factors. Show your work.

1. 6 inches to millimeters.

2. 34.62 inches to millimeters.

3. 16 1/2 inches to centimeters.

4. 48 feet to meters.

5. 90 yards to meters.

6. 27 miles to kilometers.

7. 118 mm to inches.

8. 14.5 mm to inches.

9. 2.03 m to feet.

10. 98 km to miles.

11. 150 in.² to cm²

12. 20 yd.² to m²

13. 16 pounds to kg

14. 4 tons to Mg

15. 12 kg to pounds.

16. 30 yd.³ to m³

17. 24 m³ to yd.³

18. 12 quarts to liters.

19. 25 gallons to liters.

20. 4 liters to quarts.

Convert the following by use of the conversion tables in the Appendix.

21. 3.375″ = _____ mm

22. 128 1/2″ = _____ mm

23. 63.5 mm = _____ inches

24. 37 yd.² = _____ m²

25. 11 pounds 8 ounces = _____ kg

26. 4 quarts = _____ ℓ

27. 68 °F = _____ °C

28. 3783 mm = _____ m

29. 4.2 kg = _____ g

30. 200 cℓ = _____ ℓ

2. The size of a certain classroom is 22 feet by 32 feet. Convert these measurements to meters and find the area of the floor in square meters (m²).

3. The normal house wall is framed using 1 5/8″ × 3 5/8″ × 8′ lumber for studs. Which of the following standard metric sizes of lumber most closely approximates this customary size?

   a. 22 mm × 75 mm × 1.8 m

   b. 25 mm × 75 mm × 2.1 m

   c. 40 mm × 100 mm × 2.4 m

   d. 50 mm × 100 mm × 2.4 m

EXERCISE 19-2

Solve these problems. Show your work.

1. A door opening is 6 feet 8 inches in height. What does it measure in centimeters? Meters?

4. A truck was loaded with 6528 pounds of gravel. What is the weight of the gravel in kilograms? In megagrams?

5. What is the weight of an 80 pound bag of cement in kilograms?

8. An automobile gasoline tank holds 23.5 gallons. How many liters of gasoline will it hold?

6. A materials bin (storage area) measures 14 feet wide, 24 feet long, and 8 feet deep. What is the volume of this bin in cubic meters (m³)?

9. A temperature reading of 72 °F would be what reading in degrees Celsius?

7. The oil reservoir on a machine measures 12 inches by 20 inches by 30 inches. In cubic meters (m³), what is its capacity?

10. A temperature reading of 25 °C would be what reading in degrees Fahrenheit?

PERCENTAGE

Percentage is a term frequently used in business, in industry, and in our everyday lives.

Examples: A lumber dealer advertises a 10 percent discount on fir plywood; a bank charges a welding shop 7 percent interest on a loan.

Percent means "by the hundred." It is indicated by the sign %. One percent of a number is one hundredth part of the number. One hundredth part of a number may be indicated by the common fraction 1/100, by the decimal fraction .01, or by the percent 1%.

Problem: Find 3% of 1750

$$
\begin{array}{r}
1750 \\
\underline{.03} \\
52.50
\end{array}
$$

Since percent means hundredths, 3% of 1750 would be .03 (3/100) of that number. We multiply 1750 by .03 and get 52.50.

CHANGING DECIMAL TO PERCENT

Decimals that have two places such as .25, .65, .90 (hundredths), may be changed to percents by simply removing the decimal points and adding percent signs like this:

.25 = 25%
.65 = 65%
.90 = 90%

Decimals that have one place .2, .6, .9 (tenths), may be changed to percents (hundredths) by annexing a zero and adding a percent sign:

.2 = 20%
.6 = 60%
.9 = 90%

If there are more than two decimal places such as .385, .218, .7684, the percent may be shown as a decimal mixed number:

.385 = 38.5%
.218 = 21.8%
.7684 = 76.84%

CHANGING PERCENTS TO DECIMALS

Percents such as 6%, 47%, 52%, 75%, etc., may be changed to decimals by dropping the percent signs and pointing off two decimal places:

6% = .06
47% = .47
52% = .52
75% = .75

CHANGING COMMON FRACTIONS TO PERCENTS

In changing common fractions, such as 1/5, 3/8, 13/16, etc., to percents, the first step is to change the common fractions to decimals. Then, the decimals are changed to percents.

1/5   = .20   = 20%
3/8   = .375 = 37.5%
13/16 = .8125 = 81.25%

CHANGING PERCENTS TO COMMON FRACTIONS

In changing percents to common fractions, such as 4%, 25%, 75%, etc., we first change the percents to decimals. Then, we change the decimals to common fractions and reduce to the lowest terms.

4% = .04 = 4/100 = 1/25
25% = .25 = 25/100 = 1/4
75% = .75 = 75/100 = 3/4

EXERCISE 20-1

Solve the following problems. Show your work.

1. 7% of 100 = _____

2. 13% of 185 = _____

3. 22% of 16.2 = _____

4. 25% of 16.2 = _____

5. 5.3% of 75 = _____

6. 37 1/2% of 80 = _____

7. 17% of 250 = _____

8. 28% of 380 = _____

9. 80% of 860 = _____

10. 71% of 1020 = _____

11. 52% of 150 = _____

12. 19% of 780 = _____

BASIC MATHEMATICS

13. 8% of 240 = _____

14. 32% of 384 = _____

15. 45% of 360 = _____

16. 125% of 400 = _____

17. 1/2% of 10 = _____

18. 12 1/2% of 8 = _____

19. 12% of 20 = _____

20. 16% of 1250 = _____

21. 36% of 125 = _____

22. 33 1/3% of 7425 = _____

23. 12 1/2% of 87 1/2 = _____

24. 87 1/2% of 720 = _____

25. 6% of 36 = _____

26. 41% of 140 = _____

27. 110% of 418 = _____

28. 28% of 32 = _____

29. .25% of 6 = _____

30. 24 1/2% of 136 = _____

31. 4% of 3/4 = _____

32. 500% of 6 = _____

33. 30% of 21 = _____

FINDING THE PERCENT ONE NUMBER IS OF ANOTHER

To find what percent one number is of another, divide the number for which you desire the percent by the second number.

Example: What percent is 45 of 180?

$$
\begin{array}{r}
.25 = 25\% \\
180\overline{)45.00} \\
\underline{36\ 0} \\
9\ 00 \\
\underline{9\ 00}
\end{array}
$$

Example: What percent is 56 of 448?

$$
\begin{array}{r}
.125 = 12.5\% \\
448\overline{)56.000} \\
\underline{44\ 8} \\
11\ 20 \\
\underline{8\ 96} \\
2\ 240 \\
\underline{2\ 240}
\end{array}
$$

Example: What percent is 480 of 2500?

$$.192 = 19.2\%$$
$$2500\overline{)480.000}$$
$$\underline{250\ 0}$$
$$230\ 00$$
$$\underline{225\ 00}$$
$$5\ 000$$
$$\underline{5\ 000}$$

To check the above problem:

$$2500 \times .192 = 480$$

ESTIMATING PERCENTS

It is helpful to be able to estimate the value of a percent even before multiplying. This can be done rather easily for 10%.

Example:

10% of 60 = .10 × 60 = 6.0

To find 10% of a number, simply move the decimal point one place to the left.

Example:

10% of 80 = 8.0
10% of 56 = 5.6

To find 1% of a number, move the decimal point two places to the left.

Example:

1% of 80 = .80
1% of 56 = .56

Knowing how to find these special percents quickly will help you estimate other percents such as 2% (which is double the value of 1%) and 5% (which is one half the value of 10%).

Hint:

If the percentage rate is 3%, try 3% of the number (such as 1200) which may be 36, 3600, or 3600.00. The number 1200 is 12 one-hundreds. So, 3 × 12 = 36 and 3 × 1200 = 3600. Move the decimal point to the left and your best answer is 36.

EXERCISE 20-2

In each of the following problems, three possible answers are given. Circle the answer you estimate to be correct. Then in the space below, work each problem and check your estimate.

1. A stamping machine, in making an airplane rib, wastes 17% of the sheet metal fed into it. If 540 square feet of sheet metal are used on a given job, the amount of metal wasted is 9.18 square feet, 91.8 square feet, or 918 square feet.

2. A lumber dealer allows a contractor's discount of 8%. If a contractor bought $2856 worth of lumber, the discount would amount to $2.28, $22.85, or $228.48.

3. A truck driver, hauling gravel from a river bed, estimates that 7% of the load is water. If the total load weighs 2500 pounds, then the amount of water in the load is 1.75 pounds, 17.5 pounds, or 175 pounds.

4. A contractor was assessed a 4% penalty for not completing a job in the agreed time. If the bid was $60,450 for the job, the amount of loss, because of the penalty, was $241.80, $2418.00, or $24,180.00.

5. An employee estimates 2.5 hours are spent each day in the warehouse and the rest of the 8 hour day in the shop. The 2.5 hours worked in the warehouse are 3.125%, 31.25%, or 312.5% of the work day.

6. Last year, a salesperson paid 4.5% of his yearly earnings into a retirement fund. If his total earnings were $32,500, how much was paid into the retirement fund—$1462.50, $1787.50, or $2112.50?

7. A tile layer wasted 11.2 square feet of floor tile on a job. If this was 8% of the tiles purchased for the job, then 14, 140, or 1400 square feet of tile were purchased.

8. An electronics plant produced 2,460 transistors, of which 123 were found to be defective. The percent of defective transistors was .5%, 5%, or 50%.

9. A plumbing and heating contractor listed gross earnings last year at $375,000. It is estimated that gross earnings this year will be 112% of last year's. The gross earnings this year will be $4,200, $42,000, or $420,000.

10. A certain mortar contained 70% sand. If 4.2 cubic feet of sand were used in mixing mortar for a job, then the job required .6, 6, or 60 cubic feet of mortar.

11. If 12 1/2% of a job can be completed in three hours, then 2.4, 24, or 240 hours will be required to do the job.

12. In a factory, 24 out of a total of 96,000 hours were lost because of accidents. The number of hours lost due to accidents represents .025%, .25%, or 2.5% of the total hours worked.

13. A cabinetmaker sold a radial saw for $292.50. If this were 65% of the cost of the saw, what was originally paid for it? $45.00, $450.00, or $4,500.00?

16. A manufacturing job required 1800 hours to complete. The foreman said 12 1/2% of the time was on milling machines and amounted to 2.25, 22.5, or 225 hours.

14. A painting contractor purchased $127.50 worth of paints and thinners for a job. The contractor was given a discount of 15% that amounted to $0.19, $1.91, or $19.13.

17. A worker invested $21,000 of the family's $60,000 savings in a piece of property. This represented .35%, 3.5%, or 35% of the savings?

15. When hardware items are bought in amounts of over $100.00, the dealer allows a 12% discount. What is the discount on a $586.40 order? $0.70, $7.04, or $70.37?

INTEREST

Interest is money paid for the use of money. The amount of money borrowed is the **Principal;** the percent charged per year for interest is the **Rate.**

The amount of interest to be paid depends on the time the borrowed money is kept. In figuring interest we consider a year to be 12 months, each of 30 days, or 360 days (instead of 365).

In cases where the money is repaid in less than a year, we find out what fraction of a year the borrowed money is kept. If the loan is repaid at the end of six months the interest charge is 1/2 of a full year's interest; if repaid in three months, the charge is 3/12 or 1/4 of a full year's interest. If payment is made in 45 days the charge is 45/360.

Problem: Find the interest on $2,000 for 1 year at 7%.

$2,000
.07
$140.00

Problem: Find the interest on $150 for 60 days (2 months) at 8%.

$ 150
.08
6)12.00
$2.00

We find the interest for 1 year at 8% to be $12.00. Since two months are 60/360 or 1/6 of a year, we divide the annual rate by six, which gives us $2.00, the interest charge on $150 for 60 days at 8%.

EXERCISE 21-1

Solve the following problems. Show your work.

1. Find the interest on $98.00 for 3 years 6 months at 8%.

2. Find the interest on $750.50 for 1 year at 5 1/2%.

3. Find the interest on $850.00 for 4 months at 6 1/2%.

4. Find the interest on $1,170.00 for 9 months at 5%.

5. Find the interest on $21,860.00 for 2 years at 7%.

10. Find the interest on $425.00 for 30 days at 7%.

6. Find the interest on $967.15 for 11 months at 8%.

11. Find the interest on $367.50 for 90 days at 6%.

7. Find the interest on $865.34 for 18 months at 6%.

12. Find the interest on $300.00 for 60 days at 5%.

8. Find the interest on $920.00 for 21 months at 6%.

13. Find the interest on $17,260.00 for 90 days at 9%.

9. Find the interest on $11,240.00 for 1 year at 5 1/2%.

14. Find the interest on $37.63 for 6 months at 11%.

15. Find the interest on $876.00 for 13 months at 6%.

20. Find the interest on $184.34 for 1 1/2 years at 7.6%.

16. Find the interest on $24.97 for 1 year at 6.4%.

21. Find the interest on $2040 for 2 years at 7%.

17. Find the interest on $109.98 for 2 years at 7.3%.

22. Find the interest on $15,000 for 15 months at 8%.

18. Find the interest on $50.00 for 3 years at 8.1%.

23. Find the interest on $1800 for 9 months at 7 1/2%.

19. Find the interest on $11,999.00 for 1 year at 6.5%.

24. Find the interest on $65.00 for 3 months at 8%.

THE POCKET CALCULATOR IN BASIC MATHEMATICS

The pocket calculator has become a frequently used tool in business, industry, and the home. It saves valuable time in solving mathematical problems and, if used correctly, increases the accuracy of solutions.

Fig. 22-1. This pocket calculator performs all the basic functions that will be discussed in this unit.

HOW THE CALCULATOR WORKS

The electronic pocket calculator, Fig. 22-1, is the result of a high technology race to improve our capability in communications, production, and transportation—the Space Age. Since its introduction, it has become smaller in size and more powerful in function. The electronic circuitry, which is the basis of the calculator's operation, involves microscopic integrated circuits on a piece of silicon. This miniature device is called a "chip." Depending on the capacity (number of functions) of a calculator, the number of chips vary. A chip no larger than a centimeter may contain tens of thousands of electronic components.

TERMS USED IN CALCULATOR OPERATION

Calculator arithmetic—The number system used by the calculator that is different from the real number system. An eight-digit calculator computes arithmetic on this system, rounding numbers to eight digits.

Calculator function—The mathematical function built into the calculator's circuitry. It performs, on a four-function calculator, such functions as addition, subtraction, multiplication, and division. Other common functions are square root ($\sqrt{}$) and percent (%).

Clear $\boxed{\text{C}}$ —The clear key erases or clears the contents of certain memory locations to zero.

Clear entry $\boxed{\text{CE}}$ —The clear entry key sets the number currently being keyed into zero. It is also used to correct an error.

Eight-digit calculator—Designed to accept as input, store, calculate with, and display numbers up to eight digits in length. When the answer exceeds the eight digits, the calculator will register an "E" after the last digit indicating an overflow.

Floating decimal point—The decimal point positions itself automatically to fully utilize the calculator's accuracy. An example of this on an eight-digit calculator is:

$$10 \div 3 = 3.3333333$$
$$100 \div 3 = 33.333333$$
$$1000 \div 3 = 333.33333$$

Memory—The storage in a calculator for data being processed, such as numbers, operations to be performed, and programs.

HOW TO USE THE CALCULATOR

The instructions given in this section apply to most pocket calculators. If your calculator is different, study the manufacturer's directions carefully. Practice solving problems with your calculator until you have mastered all its functions.

BASIC FUNCTIONS WITH WHOLE NUMBER PROBLEMS

ADDITION $\boxed{+}$

Clear the calculator by pushing the Clear $\boxed{C}$ key.

Problem 1: 7 + 5

Enter (push buttons):

$\boxed{7}$ $\boxed{+}$ $\boxed{5}$ $\boxed{=}$ 12 (answer shown on display)

Record the answer on paper, if desired. Push the $\boxed{C}$ key to clear calculator for next problem.

Problem 2: 386 + 414

$\boxed{3}$ $\boxed{8}$ $\boxed{6}$ $\boxed{+}$ $\boxed{4}$ $\boxed{1}$ $\boxed{4}$ $\boxed{=}$ 800

Problem 3: 9503 + 12116

$\boxed{9}$ $\boxed{5}$ $\boxed{0}$ $\boxed{3}$ $\boxed{+}$ $\boxed{1}$ $\boxed{2}$ $\boxed{1}$ $\boxed{1}$ $\boxed{6}$ $\boxed{=}$ 21619

It is always a good practice to check each entry on the display panel before entering the next function.

CORRECTING AN ENTRY $\boxed{CE}$

If you should make a wrong entry, you can correct the entry without erasing the entire problem. Most calculators erase the number back to the function (+, −, ÷, etc.) sign:

Example:
                               ERROR   CLEAR ENTRY

$\boxed{9}$ $\boxed{5}$ $\boxed{0}$ $\boxed{3}$ $\boxed{+}$ $\boxed{1}$ $\boxed{2}$ $\boxed{2}$ $\boxed{CE}$ $\boxed{1}$ $\boxed{2}$ $\boxed{1}$ $\boxed{1}$ $\boxed{6}$ $\boxed{=}$ 21619

When the wrong function sign has been entered in a problem, just enter the correct function and proceed. The CE is not necessary to clear a function entry.

Example: ERROR→   ←CORRECT ENTRY

9 5 0 3 − + 1 2 1 1 6 = 21619

Try this problem. Include the error and correct entry:

1 7 − + 6 4 CE 6 2 + 8 4 + 3 9 =

Your answer should be 202. Now try some of the larger problems on page 11 and 12 of this book. Check your answers by adding in the numbers in reverse order.

SUBTRACTION −

Clear C the calculator.

Problem 1: 43 − 19

Enter:

4 3 − 1 9 = 24

Record the answer on paper, if desired. Clear C the calculator.

Problem 2: 869 − 574

8 6 9 − 5 7 4 = 295

Problem 3: 37105 − 18548

3 7 1 0 5 − 1 8 5 4 8 = 18557

Remember, to correct a wrong entry, push the CE key and re-enter the correct numbers.

Example 1: 43226 − 36004 = 7222

4 3 2 2 6 − 3 6 0 4 CE 3 6 0 0 4 = 7222

Example 2: 82597 − 58103 = 24494

8 2 5 9 7 × − 5 8 1 0 3 = 24494

Study the procedure on page 13 for checking the answer for a subtraction problem. Before erasing the answer (remainder) to a problem on the calculator, add the subtrahend to this figure. The sum should equal the minuend.

Now try some of the problems on pages 13 to 16 of this book. Check your answers.

MULTIPLICATION ⊠

Clear C the calculator.

Problem 1: 37 × 82

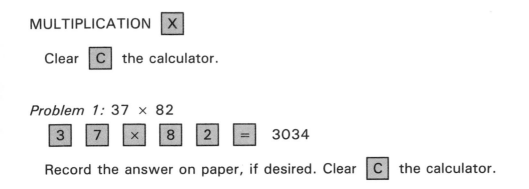

3 7 × 8 2 = 3034

Record the answer on paper, if desired. Clear C the calculator.

Problem 2: 2438 × 769 = 1850442

Check the answer for the above multiplication problem on the calculator using the procedure shown on pages 17 and 18. Or you can check your answer by dividing the product by either the multiplicand or the multiplier. The figure you get should be the opposite of the one you used to check the answer.

Example:

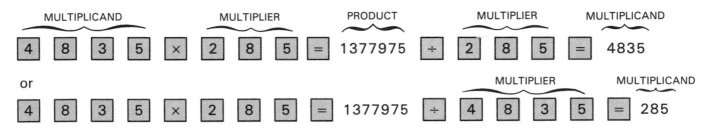

MULTIPLICAND	MULTIPLIER	PRODUCT	MULTIPLIER	MULTIPLICAND
4 8 3 5 ×	2 8 5 =	1377975 ÷	2 8 5 =	4835

or

			MULTIPLIER	MULTIPLICAND
4 8 3 5 ×	2 8 5 =	1377975 ÷	4 8 3 5 =	285

Use your calculator to solve some of the problems on pages 18 to 20.

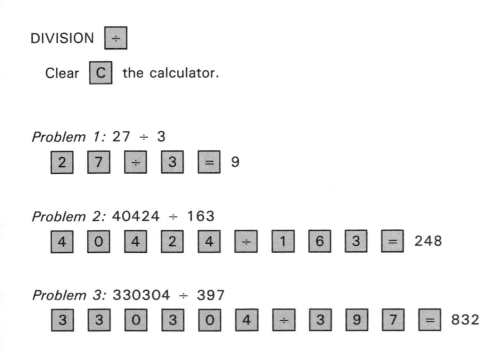

DIVISION ÷

Clear C the calculator.

Problem 1: 27 ÷ 3

2 7 ÷ 3 = 9

Problem 2: 40424 ÷ 163

4 0 4 2 4 ÷ 1 6 3 = 248

Problem 3: 330304 ÷ 397

3 3 0 3 0 4 ÷ 3 9 7 = 832

Answers to problems with remainder will be shown as decimal fractions on the calculator.

Example: 73 ÷ 3 = 24 with 1 remainder.

3 will go into 73, 24 times with 1 remainder, or 24 1/3.

The calculator will show this as a mixed decimal number:

[7] [3] [÷] [3] [=] 24.333333

Check your answers.

You may check the answers for division problems, as noted on page 21, by multiplying the quotient by the divisor.

Example 1: 19092 ÷ 37 = 516

[1] [9] [0] [9] [2] [÷] [3] [7] [=] 516 [×] [3] [7] [=] 19092

Example 2 (with decimal answer [quotient]): 482 ÷ 32 = 15.0625

[4] [8] [2] [÷] [3] [2] [=] 15.0625 [×] [3] [2] [=] 482

Now try some of the problems on pages 21 to 24.

BASIC FUNCTIONS WITH MIXED DECIMAL PROBLEMS

ADDITION [+] OF DECIMAL [.] NUMBERS

Decimal numbers are easily added with the calculator. Just enter the decimal point [.] as it appears in the number and proceed.

Problem 1: 3.375 + 4.5 + 5.25

[3] [.] [3] [7] [5] [+] [4] [.] [5] [+] [5] [.] [2] [5] [=] 13.125

Problem 2: 8.127 + 0.066 + 2.2

[8] [.] [1] [2] [7] [+] [.] [0] [6] [6] [+] [2] [.] [2] [=] 10.393

Now try some of the problems on pages 50 to 54. Check your answers.

SUBTRACTION [−] OF DECIMAL [.] NUMBERS.

Problem 1: 803.76 − 584.87

[8] [0] [3] [.] [7] [6] [−] [5] [8] [4] [.] [8] [7] [=] 218.89

Problem 2: 387.106 − 95.4

[3] [8] [7] [.] [1] [0] [6] [−] [9] [5] [.] [4] [=] 291.706

Remember to check each entry before entering the next function.

Now try some of the problems on pages 55 to 57. Check your answers.

MULTIPLICATION $\boxed{\times}$ OF DECIMAL $\boxed{.}$ NUMBERS

Problem 1: 506.14 × 37.2

$\boxed{5}$ $\boxed{0}$ $\boxed{6}$ $\boxed{.}$ $\boxed{1}$ $\boxed{4}$ $\boxed{\times}$ $\boxed{3}$ $\boxed{7}$ $\boxed{.}$ $\boxed{2}$ $\boxed{=}$ 18828.408

Problem 2: 927.06 × 0.95

$\boxed{9}$ $\boxed{2}$ $\boxed{7}$ $\boxed{.}$ $\boxed{0}$ $\boxed{6}$ $\boxed{\times}$ $\boxed{.}$ $\boxed{9}$ $\boxed{5}$ $\boxed{=}$ 880.707

Remember to check each entry before the next function.

Try some of the problems on pages 58 to 60. Check your answers.

DIVISION $\boxed{\div}$ OF DECIMAL $\boxed{.}$ NUMBERS

Problem 1: 832.0752 ÷ 47.52

$\boxed{8}$ $\boxed{3}$ $\boxed{2}$ $\boxed{.}$ $\boxed{0}$ $\boxed{7}$ $\boxed{5}$ $\boxed{2}$ $\boxed{\div}$ $\boxed{4}$ $\boxed{7}$ $\boxed{.}$ $\boxed{5}$ $\boxed{2}$ $\boxed{=}$ 17.51

Problem 2: 2500.4552 ÷ 72.08

$\boxed{2}$ $\boxed{5}$ $\boxed{0}$ $\boxed{0}$ $\boxed{.}$ $\boxed{4}$ $\boxed{5}$ $\boxed{5}$ $\boxed{2}$ $\boxed{\div}$ $\boxed{7}$ $\boxed{2}$ $\boxed{.}$ $\boxed{0}$ $\boxed{8}$ $\boxed{=}$ 34.69

Remember to check each entry before entering the next function.

Try some of the problems on pages 62 to 65. Check your answers.

CHANGING COMMON FRACTIONS TO DECIMALS USING THE CALCULATOR

You studied in Unit 17 the procedure for changing common fractions to decimals. The numerator is divided by the denominator.

The calculator performs this operation easily and quickly.

Example 1: Change $\frac{5}{8}$ to a decimal fraction.

$\boxed{C}$
$\boxed{5}$ $\boxed{\div}$ $\boxed{8}$ $\boxed{=}$ .625

Example 2: Change $\frac{13}{16}$ to a decimal fraction.

$\boxed{1}$ $\boxed{3}$ $\boxed{\div}$ $\boxed{1}$ $\boxed{6}$ $\boxed{=}$ .8125

Change the following common fractions to decimals:

a. $\dfrac{3}{32}$ b. $\dfrac{11}{64}$

c. $\dfrac{31}{32}$ d. $\dfrac{3}{1000}$

You should have the following answers:

a. 0.09375   b. 0.171875

c. 0.96875   d. 0.003

Now try some of the problems on pages 66 to 70.

SOLVING SEVERAL FUNCTIONS IN A SINGLE PROBLEM

In a problem that involves several functions, such as: 8 × (4 + (18 ÷ 3) − 5) ÷ 2 = 20, there is an order of operation. Mathematicians have agreed upon a rule for **order of operations.** Each calculation must be performed in a definite order to obtain the correct answer. The order is as follows:

1. Parentheses.
2. Powers. (The number of times a quantity is multiplied by itself: 5^3 = 5 × 5 × 5 = 125. This function is seldom used at the level of mathematics presented in this text.)
3. Multiplication and Division.
4. Addition and Subtraction.

PARENTHESES

Parentheses are always solved in pairs, beginning with the innermost pair and working out.

Example:

8 × (3 + (28 ÷ 4) − 6) ÷ 2 =
   (8 × (3 + 7 − 6) ÷ 2 =
      (8 × 4) ÷ 2 =
      32 ÷ 2 = 16

Try these problems on your calculator:

a. 10 × (4 + (16 ÷ 4) − 3) ÷ 2

b. 6 × (3 + (80 ÷ 2) − 7) ÷ 3

You should have these answers:

a. 25   b. 72

MULTIPLICATION AND DIVISION

These functions are performed as they appear, while working from left to right.

Example 1:

$$12 \div 3 \times 5 \div 4 \times 8 =$$
$$4 \times 5 \div 4 \times 8 =$$
$$20 \div 4 \times 8 =$$
$$5 \times 8 = 40$$

Example 2:

$$24 \times 4 \div 16 \div 3 \times 10 =$$
$$96 \div 16 \div 3 \times 10 =$$
$$6 \div 3 \times 10 =$$
$$2 \times 10 = 20$$

ADDITION AND SUBTRACTION

These functions are performed as they appear from left to right.

Example 1:

$$18 + 10 - 5 - 2 - 7 + 12 =$$
$$28 - 5 - 2 - 7 + 12 =$$
$$23 - 2 - 7 + 12 =$$
$$21 - 7 + 12 =$$
$$14 + 12 = 26$$

Example 2:

$$30 - 5 - 3 + 7 - 13 =$$
$$25 - 3 + 7 - 13 =$$
$$22 + 7 - 13 =$$
$$29 - 13 = 16$$

Remember, in problems involving several functions, multiplication and division operations are performed before addition and subtraction. For example, $28 - 5 \times 3$, is performed as follows:

$$28 - (5 \times 3) =$$
$$28 - 15 = 13$$

Example:

$$35 \div 7 + 6 =$$
$$5 + 6 = 11$$

For more complex problems, you may find it helpful to place parentheses around the parts to be worked in order.

Example:

$$7 \times 3 + 8 \times 4 \div 2 - 6 =$$
$$(7 \times 3) + ((8 \times 4) \div 2) - 6 =$$
$$21 + (32 \div 2) - 6 =$$
$$21 + 16 - 6 =$$
$$37 - 6 = 31$$

Note that the answer is the same for the previous example if you perform the division before the multiplication:

$$(7 \times 3) + (8) \times (4 \div 2) - 6 =$$
$$21 + (8 \times 2) - 6 =$$
$$21 + 16 - 6 =$$
$$37 - 6 = 31$$

Try these problems on your calculator:

a. $30 + 6 \times 4 - 18 \div 2$
b. $81 \div 3 - 13 \times 2 + 6$
c. $4 + 8 + 6 \times 3 - 10$

You should have these answers:

a. 45   b. 7   c. 20

TRUNCATION AND ROUNDING

When certain calculations are performed on the calculator, slight inaccuracies may occur. For example, $1 \div 3$ is displayed on an eight-digit calculator as 0.3333333.

When this displayed value is multiplied by 3, it does not return the original value of 1, but rather 0.9999999. This kind of an error is known as **Truncation.**

Another type of inaccuracy occurs when 1 is added to the next to the last digit if the last digit is 5 or more. For example, $2 \div 3$ is displayed on an eight-digit calculator as 0.6666666. Rounding changes this next to the last digit to 7, or 0.666667.

When this displayed value is multiplied by 3 it does not return the original value of 2, but rather 2.000001. This kind of an error is known as **Rounding.**

SPECIAL FUNCTIONS OF THE CALCULATOR

SQUARE ROOT $\boxed{\sqrt{}}$

Finding the **Square root** of a quantity (number) is to find a number which, when multiplied times itself, will produce the original quantity. For example, the square root of 9 is 3:

 3

Find the square root of:

a. 16   d. 12.25
b. 25   e. 81
c. 2.25   f. 289

You should have these answers:

a. 4   b. 5   c. 1.5   d. 3.5   e. 9   f. 17

THE POCKET CALCULATOR IN BASIC MATH

SQUARING A NUMBER

Squaring a number means multiplying the number times itself:

4 × 4 = 16
12 × 12 = 144

The calculator handles the squaring of a number by the following procedure. Use the number 17.

Clear  289

Try squaring these numbers:

a. 21   c. 110
b. 44   d. 345

You should have these answers:

a. 441   b. 1936   c. 12100   d. 119025

PERCENTAGE %

Percentage problems are easily solved with calculators having a percentage % function.

To find the percentage of a number, key in the number and multiply by the percent.

Example: 80 × 25%

DISPLAY

Example 2: 6% of $160

Problem:

Find the price of a portable power tool that has been discounted. It sells regularly for $210. It is on sale at a 15% discount. What is the sales price including 6% tax?

Key in the regular price of $210. Find the amount of the discount and subtract. Add the 6% sales tax.

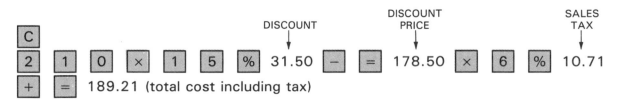

189.21 (total cost including tax)

Try some of the problems on pages 81 to 86.

INTEREST

Interest is the money paid for the use of money. The rate of interest is the percent charged for a period of time—usually a year. Interest is calculated in the same manner as percentage. For example, suppose you have $200 borrowed or owed for items purchased. There is an 8% rate of interest per year (percentage). To solve for the amount of interest due:

DISPLAY → 16   AMOUNT DUE INCLUDING INTEREST → 216

2 0 0 × 8 % 16 + = 216

Problem:

Find the amount of interest due at the end of 3 months on the following loan:

$8.50 amount borrowed for 3 months (3 ÷ 12).

7% rate for 12 months.

C

8 5 0 × 7 % DISPLAY → 59.5 × 3 ÷ 1 2 = DISPLAY → 14.875

Answer: $14.88 interest due at end of 3 months.

Try some of the problems on pages 87 to 89.

FOUR-KEY MEMORY CM RM M− M+

CM Clear Memory.

RM Recall Memory, sometimes labeled MR .

M+ Adds the display number (in temporary storage) to M and places the answer in M.

M− Subtracts the display number (in temporary storage) from M and places the answer in M.

Your calculator may have the four-key memory that is useful in performing multiple function problems. A four-key memory system calculator has an extra memory location that we call M. This memory can store a number for future use in solving problems.

For example, 82 × 4 − 17 × 5 can be solved with the use of the four-key memory without having to record a partial answer on paper. Follow this procedure:

Clear C the calculator.

Clear CM the calculator memory storage.

Enter the first part of the above problem:

8 2 × 4 = 328 (shown on display and in temporary storage)

Enter M+ to store in memory (a positive number).

Clear C the display panel and enter second part of problem:

 85 (shown on display panel)

Enter M– to subtract from memory (a negative number).

Enter RM to obtain the difference 243 (shown on display).

Most calculators will display a small "M" along with the stored number, indicating the calculator memory is being used.

Problem: 128 × 3 – 160 ÷ 4

[1] [2] [8] [×] [3] [=] 384

Enter M+ to store in memory

Clear C

Enter second part:

[1] [6] [0] [÷] [4] [=] 40

Enter M– to store in memory.

Enter RM to obtain the difference, 344.

Solve these problems using the memory:

a. 68 × 8 – 55 × 4
b. 250 × 5 – 810 ÷ 9
c. 96 × 4 – 327 × 2

You should have these answers:

a. 324   b. 1160   c. 1038

To store in memory, part of a problem to be multiplied or divided by another part of the problem, follow this procedure:

Problem 1: 52 × 20 × 30 ÷ 2

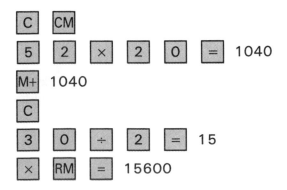

Problem 2: 48 ÷ 2 × 16 × 5

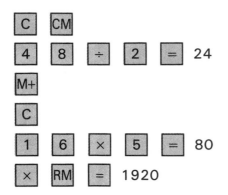

Solve these problems:

a. (26 × 8) × (80 ÷ 4)
b. (34 × 6) ÷ (12 × 4)

You should have these answers:

a. 4160   b. 4.25

CONSTANTS

Your calculator, with a memory, can be used to store a **Constant** (a uniform or unchanging number) such as π (pi) (3.1416) used to calculate measures of circles.

Formulas for circles include: a = area; c = circumference; d = diameter; r = radius; area = πr^2; circumference = $2\pi r$; diameter = $\frac{c}{\pi}$.

To enter the constant (pi) into the memory, use this procedure:

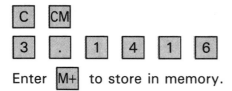

Enter [M+] to store in memory.

The constant π is now stored in the calculator's memory.

The following circle diameters are given: a = 3 inches; b = 4 inches; c = 5 inches; d = 7 inches; e = 16 inches. Using the diameter for Circle "a," we will solve Problems 1 and 2. You should continue by solving each problem for b through e.

Problem 1: Calculate the area (area = πr^2) for Circle "a."

Enter the constant (3.1416) into memory.

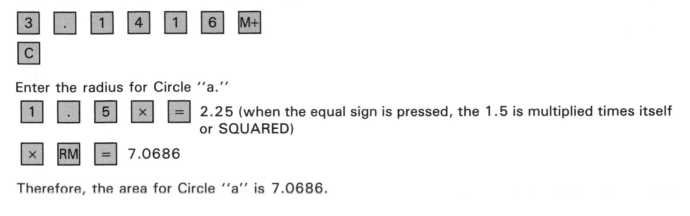

Enter the radius for Circle "a."

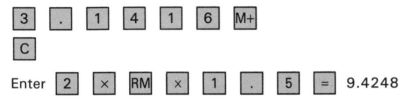

2.25 (when the equal sign is pressed, the 1.5 is multiplied times itself or SQUARED)

7.0686

Therefore, the area for Circle "a" is 7.0686.

Problem 2: Calculate the circumference (c = 2πr) for Circle "a."

Enter the constant into the memory.

Enter 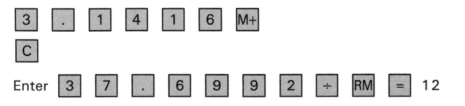 9.4248

Therefore, the circumference of Circle "a" is 9.4248.

The following circumferences are given: a = 37.6992; b = 23.562; c = 13.3518; d = 65.9736; e = 15.708. Using the circumference for Circle "a," we will solve Problem 3. You should continue by solving Problem 3 for b through e.

Problem 3: Calculate the diameter (d = $\frac{c}{\pi}$) for Circle "a."

Enter the constant into memory.

Enter  12

Therefore, the diameter for Circle "a" is 12.

You should have the following answers for the problems on constants:

1. a. 7.0686 b. 12.5664 c. 19.635 d. 38.4846 e. 201.0624
2. a. 9.4248 b. 12.5664 c. 15.708 d. 21.9912 e. 50.2656
3. a. 12 b. 7.5 c. 4.25 d. 21 e. 5

The pocket calculator is a valuable tool. Learn to use it skillfully.

MEASUREMENT, READING A RULE

READING THE FRACTIONAL-INCH RULE

In the lab, classroom, or on the job, the basic measuring tool is the rule. A rule is divided into equal parts or inches. Each inch is divided into equal fractional parts of an inch. The fractional parts are halves (1/2), quarters (1/4), eighths (1/8), and sixteenths (1/16). Some rules (metal) have graduations or divisions as small as thirty-seconds (1/32), and sixty-fourths (1/64). The denominators of the fractions. . .1/2, 1/4, etc., indicate the number of like spaces of that size, which are in a whole inch. See Figs. 23-1 and 23-2.

The drawings, Figs. 23-3 and 23-4, (with sizes enlarged) show an inch divided into halves, fourths, eighths, and sixteenths. Pay particular attention to Fig. 23-4. Note that the 1 inch line is the longest, the 1/2 inch line is next in length, and so on down to the line representing 1/16 inch that is the shortest. The number on the end of the rule, such as "16" in Figs. 23-3 and 23-4, indicates the number of divisions per inch on that rule.

In starting on this Unit you may count the spaces. After some practice this should not be necessary.

Fractional measurements are always reduced to the lowest terms. A measurement of 12/16 would be read as 3/4; 4/16 as 1/4; etc.

Fig. 23-1. Six-inch rule divided into 16ths and 8ths.

Fig. 23-2. Six-inch rule divided into 32nds and 64ths.

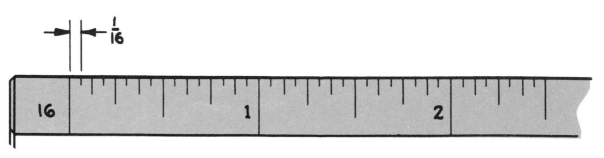

Fig. 23-3. Rule divided into 16ths.

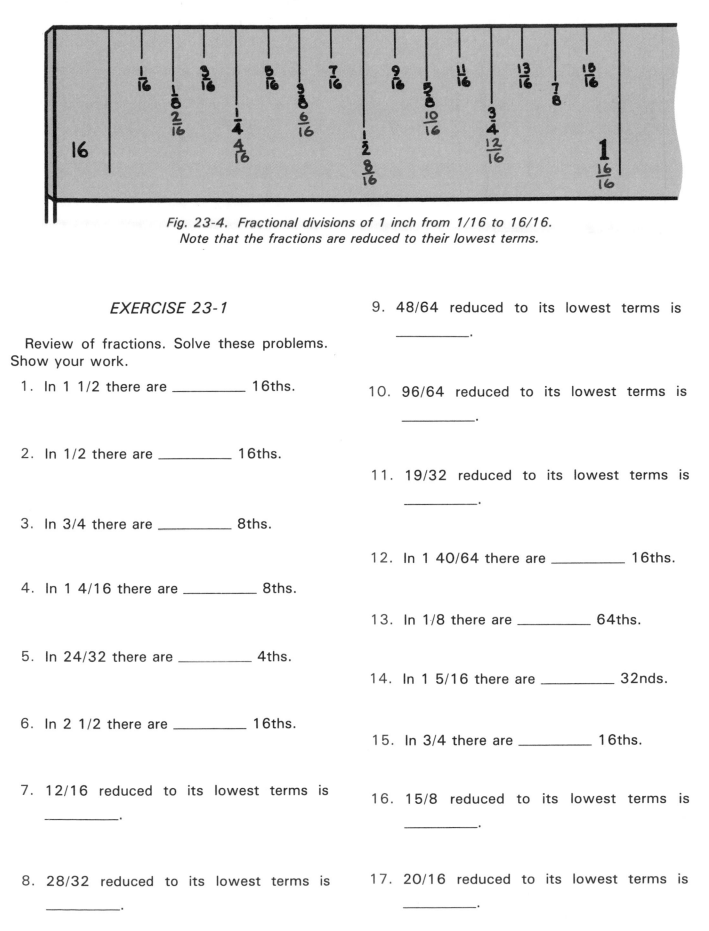

Fig. 23-4. Fractional divisions of 1 inch from 1/16 to 16/16.
Note that the fractions are reduced to their lowest terms.

EXERCISE 23-1

Review of fractions. Solve these problems.
Show your work.

1. In 1 1/2 there are _____ 16ths.

2. In 1/2 there are _____ 16ths.

3. In 3/4 there are _____ 8ths.

4. In 1 4/16 there are _____ 8ths.

5. In 24/32 there are _____ 4ths.

6. In 2 1/2 there are _____ 16ths.

7. 12/16 reduced to its lowest terms is
_____.

8. 28/32 reduced to its lowest terms is
_____.

9. 48/64 reduced to its lowest terms is
_____.

10. 96/64 reduced to its lowest terms is
_____.

11. 19/32 reduced to its lowest terms is
_____.

12. In 1 40/64 there are _____ 16ths.

13. In 1/8 there are _____ 64ths.

14. In 1 5/16 there are _____ 32nds.

15. In 3/4 there are _____ 16ths.

16. 15/8 reduced to its lowest terms is
_____.

17. 20/16 reduced to its lowest terms is
_____.

EXERCISE 23-2

The best way to learn to read a rule is to practice until you become proficient.

Note how the measurement for line (a) is filled in, then measure each line with a rule and indicate your measurements in the space provided for lines (b) to (t) inclusive.

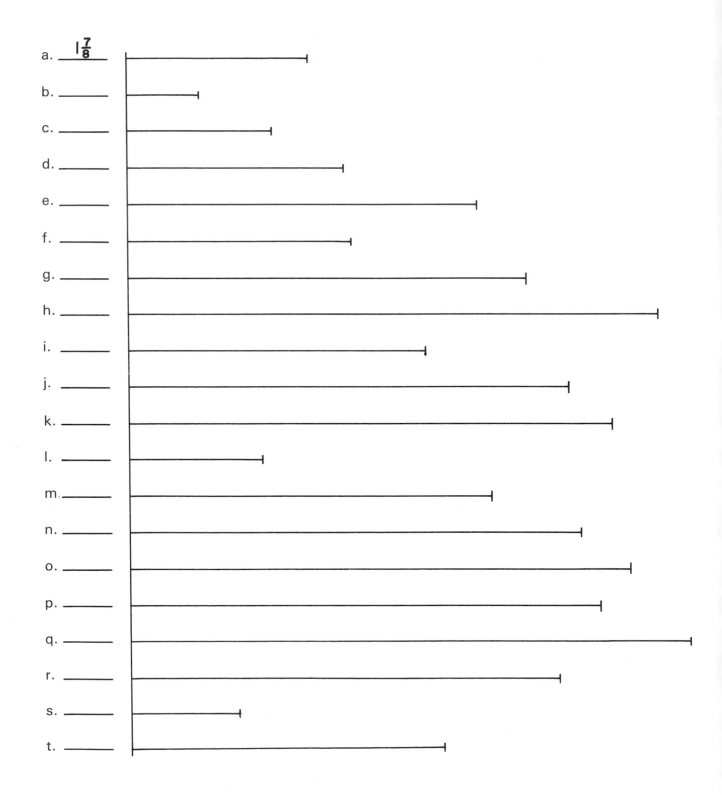

a. $1\frac{7}{8}$

b. _____

c. _____

d. _____

e. _____

f. _____

g. _____

h. _____

i. _____

j. _____

k. _____

l. _____

m. _____

n. _____

o. _____

p. _____

q. _____

r. _____

s. _____

t. _____

EXERCISE 23-3

To obtain some additional experience in measuring, use the fractional-inch rule to measure each of the lines of the template. Place your answers in the spaces at the right.

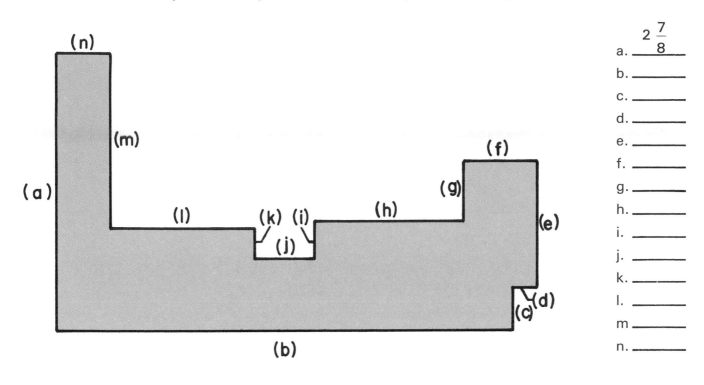

a. $2\frac{7}{8}$
b. _____
c. _____
d. _____
e. _____
f. _____
g. _____
h. _____
i. _____
j. _____
k. _____
l. _____
m _____
n. _____

EXERCISE 23-4

Rule reading experience. Note how the measurement (a) is filled in, then fill in spaces (b) to (q) inclusive.

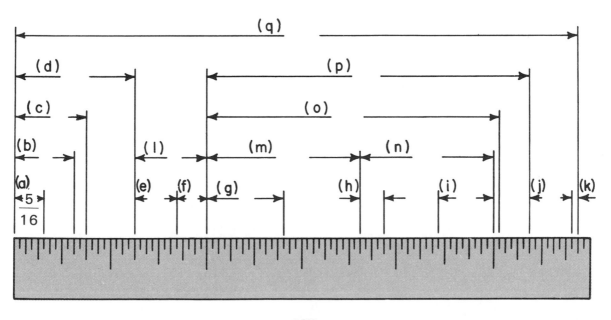

READING THE DECIMAL-INCH RULE

The decimal-inch rule is used in many industries where closer measurements are required than those possible with the fractional-inch rule.

An enlarged view of a decimal-inch rule is shown below. The "100" at the end of the rule means the inch is divided into 100 parts and that each small division is 1/100 (.01) of an inch in size. To read the decimal rule divided into 100ths of an inch, follow these steps:

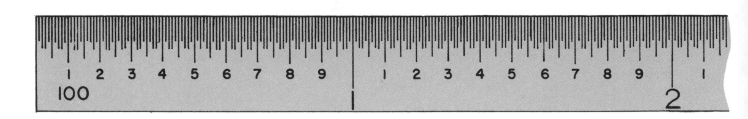

1. Study the major divisions that are divided into tenths of an inch: 1/10, 2/10, 3/10. . .and on to the 1 inch mark that represents 10/10 or 1 inch.

2. Each major division of 1/10 also represents 10/100ths (.10) of an inch; therefore, these major divisions may be read .20, .30, .40, etc.

3. For a reading of .43, start at the zero, move to the major division marked 4 and count three additional lines, or small divisions beyond. This mark represents .43 inch.

EXERCISE 23-5

Complete the readings called for in the illustration below by placing your answers in the spaces provided.

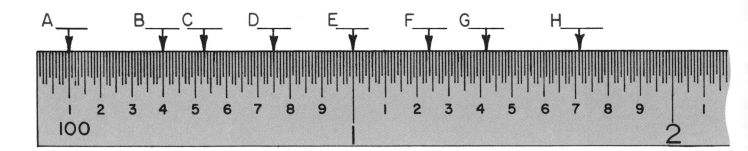

READING THE METRIC RULE

Metric rules are used with blueprints and jobs on which the dimensions are given in metric units. A metric rule is shown below.

Metric rules are read in the same way as decimal-inch rules only the units have different names. To read the metric rule, follow these steps:

1. Study the major divisions that are divided into centimeters: 1, 2, 3, and on to 15 centimeters.

2. Each centimeter is divided into 10 millimeters along one edge of the rule and into 1/2 millimeters (20 divisions) on the other.

3. For a reading of 37 millimeters, start at the zero, move to the major division marked 3, on to the mid-division mark representing 35, and count two additional divisions that represent 37 millimeters.

EXERCISE 23-6

Complete the readings called for in the illustration below by placing your answers in the spaces provided.

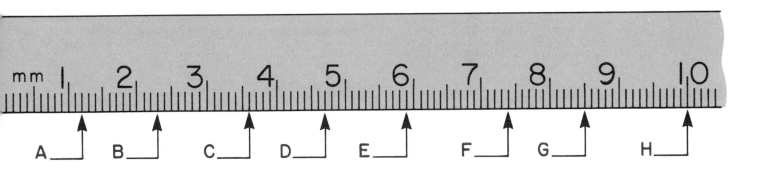

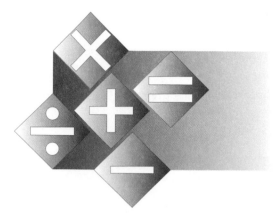

PRECISION MEASUREMENT WITH MICROMETER

In making precision measurements, the tool commonly used is the **Micrometer caliper,** or ''mike'' as it is frequently called. See Fig. 24-1. Note particularly the names of the principal micrometer parts, as indicated on the photo.

When using a micrometer of the type shown in Fig. 24-1 (outside micrometer), the material or object to be measured is placed between the **Anvil** and **Spindle faces.** The **Thimble** is then rotated (spindle moves) until the anvil and spindle contact the work. This type of micrometer measures in thousandths of an inch.

PRINCIPLE OF MICROMETER OPERATION

Accurately made screw threads on the spindle of the micrometer (spindle is attached to the thimble) rotate in a fixed nut. There are 40 threads per inch, so one complete revolution of the thimble and the spindle causes the spindle to move in or out precisely 1/40 (.025) inch.

In referring again to Fig. 24-1, you will note there is a line on the sleeve that runs length wise. This line, called the **Index line,** is divided into 40 equal parts by vertical lines, one for each revolu-

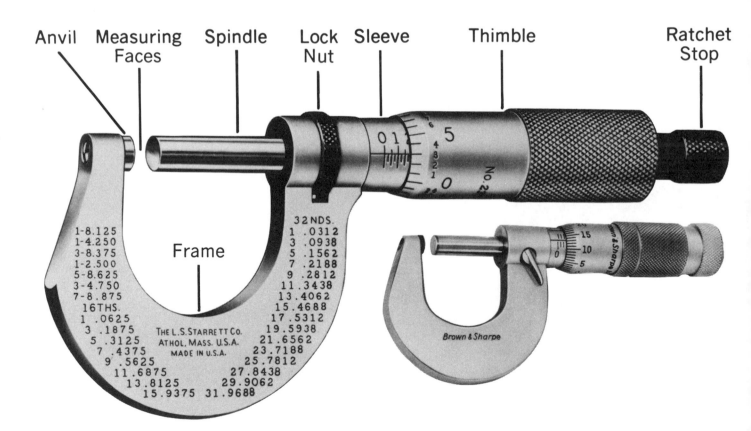

Anvil   Measuring Faces   Spindle   Lock Nut   Sleeve   Thimble   Ratchet Stop

Frame

32 NDS.

1	.0312
3	.0938
5	.1562
7	.2188
9	.2812
11	.3438
13	.4062
15	.4688
17	.5312
19	.5938
21	.6562
23	.7188
25	.7812
27	.8438
29	.9062
31	.9688

1-8.125
1-4.250
3-8.375
1-2.500
5-8.625
3-4.750
7-8.875
16THS.
1 .0625
3 .1875
5 .3125
7 .4375
9 .5625
11 .6875
13 .8125
15 .9375

THE L.S.STARRETT Co.
ATHOL, MASS. U.S.A.
MADE IN U.S.A.

Brown & Sharpe

Fig. 24-1. Above. Inch micrometer with principal parts identified. Below. Metric micrometer.

tion of the thimble. Each vertical line represents a measurement of .025 inch. Every 4th line (that is longer than the others) is numbered and represents a measurement of hundreds of thousands. The line marked "1" represents .100 inch, the line marked "2" represents .200, etc. The fifth line (one line past the numbered line "1") represents .125 (.100 + .025) inch, the sixth line represents .150 (.100 + .050), etc. The 40th line (marked "10") represents 1.000 or 1 inch (40 × .025 = 1.000).

To obtain measurements finer than .025 (in steps of .001), we make use of the numbers on the beveled edge of the thimble. The thimble is divided into 25 equal parts. Rotating the thimble from one line to the next past the index line on the sleeve, moves the spindle in or out 1/25 of the .025 it moves in one complete revolution, or .001 inch.

READING THE INCH MICROMETER

In reading a micrometer, the reading is taken at the edge of the thimble. We note the number of vertical lines visible on the sleeve between the edge of the thimble and the 0 (the 0 is not counted).

In referring to Example 1, we find there are two vertical lines visible. Since each vertical line on the sleeve represents a measurement of .025, we multiply the .025 by 2, which gives us a reading of .050.

Example 1:

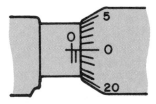

In Example 2, line marked 2 is visible, plus three additional lines. In a previous paragraph we learned that line 2 represents a measurement of .200. We add to this .025 for each of the three shorter lines visible, giving us a reading of .275.

Example 2:

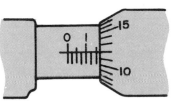

To get the mike reading for Example 3, we add:

.100 — line marked 1
.050 — 2 extra lines visible
.012 — indicated on beveled edge of thimble
.162 — reading for Example 3

Example 3:

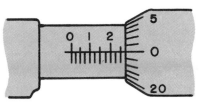

An easy way to remember the values of the various units used in reading a mike is to think of them as U.S. money: Count the figures on the sleeve. . .1, 2, etc., as dollars, the extra vertical lines on the sleeve as quarters, and the divisions on the thimble as pennies. Add up your money, then put a decimal point in front of the figure (mike reading) instead of a dollar sign.

EXERCISE 24-1

Record the readings in the space provided for
the following inch micrometer settings:

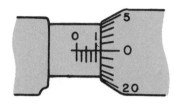

24-1-a - Reading _____

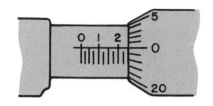

24-1-f - Reading _____

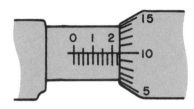

24-1-b - Reading _____

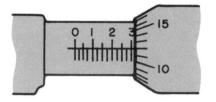

24-1-g - Reading _____

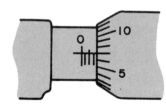

24-1-c - Reading _____

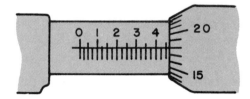

24-1-h - Reading _____

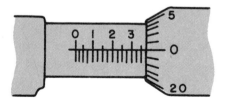

24-1-d - Reading _____

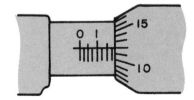

24-1-i - Reading _____

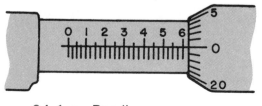

24-1-e - Reading _____

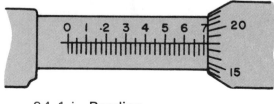

24-1-j - Reading _____

EXERCISE 24-2

Show the micrometer setting for each of the "readings" given in Problems B to H below. Start by indicating the "odd" thousandths (those beyond the last full .025) on the thimble, as shown in Problem A (.187 − .175 = .012). The thimble is numbered with the 10 and 15 to show a .012 as the setting. These "odd" thousandths will also show as part of a space on the sleeve next to the thimble. This spacing will be your first mark on the sleeve, and for Problem A represents approximately one half space (.012 is about one half of .025). The .175 is drawn on the sleeve freehand in Problem A. Remember there is one space for each .025, that each fourth line is longer, and the space between two longer lines is .100 inch.

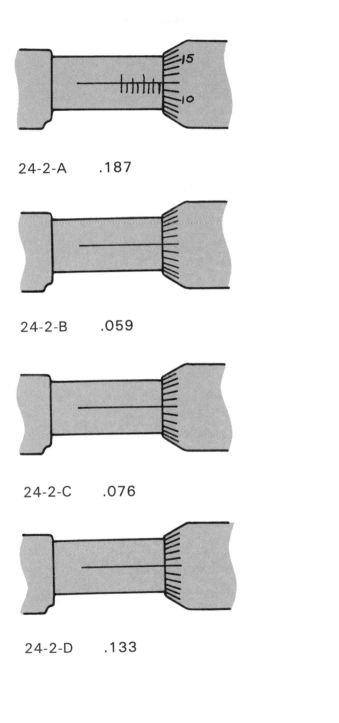

24-2-A   .187

24-2-B   .059

24-2-C   .076

24-2-D   .133

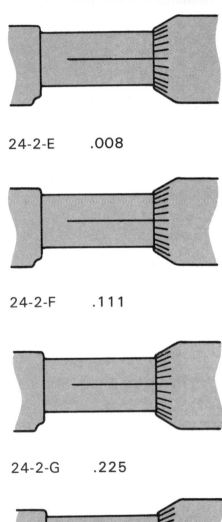

24-2-E   .008

24-2-F   .111

24-2-G   .225

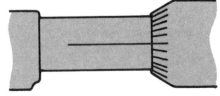

24-2-H   .102

READING THE METRIC MICROMETER

Reading a metric micrometer that measures in hundredths of a millimeter (0.01 mm) is quite similar to reading the inch micrometer—only the units are different.

The screw on the metric micrometer advances 1/2 millimeter per revolution of the thimble so two revolutions move the spindle 1 mm.

The sleeve of the micrometer is graduated in millimeters below the datum line (see below) and in half millimeters above the line. The thimble is marked in fifty divisions so that each small division represents 1/50 of 1/2 mm (one revolution of thimble) that equals 1/100 (0.01) mm.

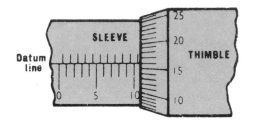

To read the metric micrometer, follow these steps:

1. Note the number of whole millimeter divisions (below datum line) on the sleeve.

2. Note whether there is a half millimeter division (above datum line) visible between the whole millimeter division and the thimble.

3. Finally, read the thimble for hundredths (division on thimble aligning with datum line).

In our example above:

Whole millimeter Divisions	= 10 × 1 mm	= 10.00 mm
Half millimeter Divisions	= 1 × 0.50 mm	= 0.50 mm
Reading on Thimble	= 16 × 0.01 mm	= 0.16 mm

Reading = 10.66 mm

EXERCISE 24-3

Record the readings in the spaces provided for the following micrometer settings.

24-3-a - Reading _____

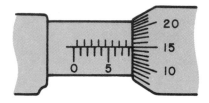

24-3-b - Reading _____

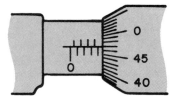

24-3-c - Reading _____

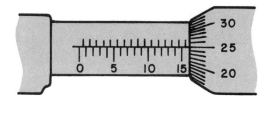

24-3-d - Reading _____

114

24-3-e - Reading _____

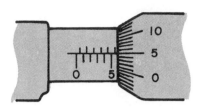

24-3-h - Reading _____

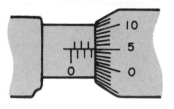

24-3-f - Reading _____

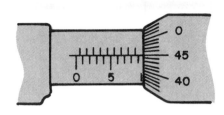

24-3-i - Reading _____

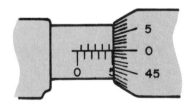

24-3-g - Reading _____

24-3-j - Reading _____

| INCHES | | MILLI-METERS | INCHES | | MILLI-METERS |
FRACTIONS	DECIMALS		FRACTIONS	DECIMALS	
	.00394	.1	15/32	.46875	11.9063
	.00787	.2		.47244	12.00
	.01181	.3	31/64	.484375	12.3031
1/64	.015625	.3969	1/2	.5000	12.70
	.01575	.4		.51181	13.00
	.01969	.5	33/64	.515625	13.0969
	.02362	.6	17/32	.53125	13.4938
	.02756	.7	35/64	.546875	13.8907
1/32	.03125	.7938		.55118	14.00
	.0315	.8	9/16	.5625	14.2875
	.03543	.9	37/64	.578125	14.6844
	.03937	1.00		.59055	15.00
3/64	.046875	1.1906	19/32	.59375	15.0813
1/16	.0625	1.5875	39/64	.609375	15.4782
5/64	.078125	1.9844	5/8	.625	15.875
	.07874	2.00		.62992	16.00
3/32	.09375	2.3813	41/64	.640625	16.2719
7/64	.109375	2.7781	21/32	.65625	16.6688
	.11811	3.00		.66929	17.00
1/8	.125	3.175	43/64	.671875	17.0657
9/64	.140625	3.5719	11/16	.6875	17.4625
5/32	.15625	3.9688	45/64	.703125	17.8594
	.15748	4.00		.70866	18.00
11/64	.171875	4.3656	23/32	.71875	18.2563
3/16	.1875	4.7625	47/64	.734375	18.6532
	.19685	5.00		.74803	19.00
13/64	.203125	5.1594	3/4	.7500	19.05
7/32	.21875	5.5563	49/64	.765625	19.4469
15/64	.234375	5.9531	25/32	.78125	19.8438
	.23622	6.00		.7874	20.00
1/4	.2500	6.35	51/64	.796875	20.2407
17/64	.265625	6.7469	13/16	.8125	20.6375
	.27559	7.00		.82677	21.00
9/32	.28125	7.1438	53/64	.828125	21.0344
19/64	.296875	7.5406	27/32	.84375	21.4313
5/16	.3125	7.9375	55/64	.859375	21.8282
	.31496	8.00		.86614	22.00
21/64	.328125	8.3344	7/8	.875	22.225
11/32	.34375	8.7313	57/64	.890625	22.6219
	.35433	9.00		.90551	23.00
23/64	.359375	9.1281	29/32	.90625	23.0188
3/8	.375	9.525	59/64	.921875	23.4157
25/64	.390625	9.9219	15/16	.9375	23.8125
	.3937	10.00		.94488	24.00
13/32	.40625	10.3188	61/64	.953125	24.2094
27/64	.421875	10.7156	31/32	.96875	24.6063
	.43307	11.00		.98425	25.00
7/16	.4375	11.1125	63/64	.984375	25.0032
29/64	.453125	11.5094	1	1.0000	25.4001

APPENDIX

APPROVED METRIC PREFIXES

Decimal Equivalent	Prefix	Symbol	Pronunciation
1 000 000 000 000	tera	T	ter'a
1 000 000 000	giga	G	ji'ga
1 000 000	mega	M*	meg'a
1 000	kilo	k*	kil'o
100	hecto	h	hek'to
10	deka	da	dek'a
0.1	deci	d	des'i
0.01	centi	c*	sen'ti
0.001	milli	m*	mil'i
0.000 001	micro	μ *	mi'kro
0.000 000 001	nano	n	nan'o
0.000 000 000 001	pico	p	pe'ko
0.000 000 000 000 001	femto	f	fem'to
0.000 000 000 000 000 001	atto	a	at'to

*Units most commonly used

CUSTOMARY SYSTEM CONVERSION FACTORS

LENGTH

1 foot (ft.) = 12 inches (in.)
1 yard (yd.) = 3 feet (ft.)
1 fathom = 6 feet (ft.)
1 rod = 5 1/2 yards (yd.)
1 mile = 1760 yards = 5280 feet

AREA

1 square foot (ft.^{2}) = 144 square inches (in.^{2})
1 square yard (yd.^{2}) = 9 square feet (ft.^{2})
1 acre = 43,560 square feet (ft.^{2})
1 square mile = 640 acres

WEIGHT (MASS)

1 pound (lb.) = 16 ounces (oz.)
1 ton = 2000 pounds (lb.)

VOLUME (DRY MEASURE)

1 cubic foot (ft.^{3}) = 1728 cubic inches (in.^{3})
1 cubic yard (yd.^{3}) = 27 cubic feet (ft.^{3})

VOLUME (LIQUID MEASURE)

1 pint (pt.) = 16 fluid ounces (oz.)
1 quart (qt.) = 2 pints (pt.)
1 gallon (gal.) = 4 quarts (qt.) or 231 cubic inches (in.^{3})

TEMPERATURE CONVERSIONS

Celsius and Fahrenheit

°C	-17.78	-6.67	0	10	15.56	20	25	26.67	37	37.78	100
°F	0	20	32	50	60	68	77	80	98.6	100	212

INCHES TO MILLIMETERS

INCREMENTS OF 100 INCHES   100 TO 900 INCHES

100	200	300	400	500	600	700	800	900
2540.0000	5080.0000	7620.0000	10160.0000	12700.0000	15240.0000	17780.0000	20320.0000	22860.0000

INCREMENTS OF 1 INCH   1 TO 109 INCHES

	0	1	2	3	4	5	6	7	8	9
0	0	25.4000	50.8000	76.2000	101.6000	127.0000	152.4000	177.8000	203.2000	228.6000
10	254.0000	279.4000	304.8000	330.2000	355.6000	381.0000	406.4000	431.8000	457.2000	482.6000
20	508.0000	533.4000	558.8000	584.2000	609.6000	635.0000	660.4000	685.8000	711.2000	736.6000
30	762.0000	787.4000	812.8000	838.2000	863.6000	889.0000	914.4000	939.8000	965.2000	990.6000
40	1016.0000	1041.4000	1066.8000	1092.2000	1117.6000	1143.0000	1168.4000	1193.8000	1219.2000	1244.6000
50	1270.0000	1295.4000	1320.8000	1346.2000	1371.6000	1397.0000	1422.4000	1447.8000	1473.2000	1498.6000
60	1524.0000	1549.4000	1574.8000	1600.2000	1625.6000	1651.0000	1676.4000	1701.8000	1727.2000	1752.6000
70	1778.0000	1803.4000	1828.8000	1854.2000	1879.6000	1905.0000	1930.4000	1955.8000	1981.2000	2006.6000
80	2032.0000	2057.4000	2082.8000	2108.2000	2133.6000	2159.0000	2184.4000	2209.8000	2235.2000	2260.6000
90	2286.0000	2311.4000	2336.8000	2362.2000	2387.6000	2413.0000	2438.4000	2463.8000	2489.2000	2514.6000
100	2540.0000	2565.4000	2590.8000	2616.2000	2641.6000	2667.0000	2692.4000	2717.8000	2743.2000	2768.6000

INCREMENTS OF .001 INCH   .001 TO 1.009 INCHES

	.000	.001	.002	.003	.004	.005	.006	.007	.008	.009
0	0	.0254	.0508	.0762	.1016	.1270	.1524	.1778	.2032	.2286
.01	.2540	.2794	.3048	.3302	.3556	.3810	.4064	.4318	.4572	.4826
.02	.5080	.5334	.5588	.5842	.6096	.6350	.6604	.6858	.7112	.7366
.03	.7620	.7874	.8128	.8382	.8636	.8890	.9144	.9398	.9652	.9906
.04	1.0160	1.0414	1.0668	1.0922	1.1176	1.1430	1.1684	1.1938	1.2192	1.2446
.05	1.2700	1.2954	1.3208	1.3462	1.3716	1.3970	1.4224	1.4478	1.4732	1.4986
.06	1.5240	1.5494	1.5748	1.6002	1.6256	1.6510	1.6764	1.7018	1.7272	1.7526
.07	1.7780	1.8034	1.8288	1.8542	1.8796	1.9050	1.9304	1.9558	1.9812	2.0066
.08	2.0320	2.0574	2.0828	2.1082	2.1336	2.1590	2.1844	2.2098	2.2352	2.2606
.09	2.2860	2.3114	2.3368	2.3622	2.3876	2.4130	2.4384	2.4638	2.4892	2.5146
.10	2.5400	2.5654	2.5908	2.6162	2.6416	2.6670	2.6924	2.7178	2.7432	2.7686
.11	2.7940	2.8194	2.8448	2.8702	2.8956	2.9210	2.9464	2.9718	2.9972	3.0226
.12	3.0480	3.0734	3.0988	3.1242	3.1496	3.1750	3.2004	3.2258	3.2512	3.2766
.13	3.3020	3.3274	3.3528	3.3782	3.4036	3.4290	3.4544	3.4798	3.5052	3.5306
.14	3.5560	3.5814	3.6068	3.6322	3.6576	3.6830	3.7084	3.7338	3.7592	3.7846
.15	3.8100	3.8354	3.8608	3.8862	3.9116	3.9370	3.9624	3.9878	4.0132	4.0386
.16	4.0640	4.0894	4.1148	4.1402	4.1656	4.1910	4.2164	4.2418	4.2672	4.2926
.17	4.3180	4.3434	4.3688	4.3942	4.4196	4.4450	4.4704	4.4958	4.5212	4.5466
.18	4.5720	4.5974	4.6228	4.6482	4.6736	4.6990	4.7244	4.7498	4.7752	4.8006
.19	4.8260	4.8514	4.8768	4.9022	4.9276	4.9530	4.9784	5.0038	5.0292	5.0546
.20	5.0800	5.1054	5.1308	5.1562	5.1816	5.2070	5.2324	5.2578	5.2832	5.3086
.21	5.3340	5.3594	5.3848	5.4102	5.4356	5.4610	5.4864	5.5118	5.5372	5.5626
.22	5.5880	5.6134	5.6388	5.6642	5.6896	5.7150	5.7404	5.7658	5.7912	5.8166
.23	5.8420	5.8674	5.8928	5.9182	5.9436	5.9690	5.9944	6.0198	6.0452	6.0706
.24	6.0960	6.1214	6.1468	6.1722	6.1976	6.2230	6.2484	6.2738	6.2992	6.3246
.25	6.3500	6.3754	6.4008	6.4262	6.4516	6.4770	6.5024	6.5278	6.5532	6.5786
.26	6.6040	6.6294	6.6548	6.6802	6.7056	6.7310	6.7564	6.7818	6.8072	6.8326
.27	6.8580	6.8834	6.9088	6.9342	6.9596	6.9850	7.0104	7.0358	7.0612	7.0866
.28	7.1120	7.1374	7.1628	7.1882	7.2136	7.2390	7.2644	7.2898	7.3152	7.3406
.29	7.3660	7.3914	7.4168	7.4422	7.4676	7.4930	7.5184	7.5438	7.5692	7.5946
.30	7.6200	7.6454	7.6708	7.6962	7.7216	7.7470	7.7724	7.7978	7.8232	7.8486
.31	7.8740	7.8994	7.9248	7.9502	7.9756	8.0010	8.0264	8.0518	8.0772	8.1026
.32	8.1280	8.1534	8.1788	8.2042	8.2296	8.2550	8.2804	8.3058	8.3312	8.3566
.33	8.3820	8.4074	8.4328	8.4582	8.4836	8.5090	8.5344	8.5598	8.5852	8.6106
.34	8.6360	8.6614	8.6868	8.7122	8.7376	8.7630	8.7884	8.8138	8.8392	8.8646
.35	8.8900	8.9154	8.9408	8.9662	8.9916	9.0170	9.0424	9.0678	9.0932	9.1186
.36	9.1440	9.1694	9.1948	9.2202	9.2456	9.2710	9.2964	9.3218	9.3472	9.3726
.37	9.3980	9.4234	9.4488	9.4742	9.4996	9.5250	9.5504	9.5758	9.6012	9.6266
.38	9.6520	9.6774	9.7028	9.7282	9.7536	9.7790	9.8044	9.8298	9.8552	9.8806
.39	9.9060	9.9314	9.9568	9.9822	10.0076	10.0330	10.0584	10.0838	10.1092	10.1346
.40	10.1600	10.1854	10.2108	10.2362	10.2616	10.2870	10.3124	10.3378	10.3632	10.3886
.41	10.4140	10.4394	10.4648	10.4902	10.5156	10.5410	10.5664	10.5918	10.6172	10.6426
.42	10.6680	10.6934	10.7188	10.7442	10.7696	10.7950	10.8204	10.8458	10.8712	10.8966
.43	10.9220	10.9474	10.9728	10.9982	11.0236	11.0490	11.0744	11.0998	11.1252	11.1506
.44	11.1760	11.2014	11.2268	11.2522	11.2776	11.3030	11.3284	11.3538	11.3792	11.4046
.45	11.4300	11.4554	11.4808	11.5062	11.5316	11.5570	11.5824	11.6078	11.6332	11.6586
.46	11.6840	11.7094	11.7348	11.7602	11.7856	11.8110	11.8364	11.8618	11.8872	11.9126
.47	11.9380	11.9634	11.9888	12.0142	12.0396	12.0650	12.0904	12.1158	12.1412	12.1666
.48	12.1920	12.2174	12.2428	12.2682	12.2936	12.3190	12.3444	12.3698	12.3952	12.4206
.49	12.4460	12.4714	12.4968	12.5222	12.5476	12.5730	12.5984	12.6238	12.6492	12.6746
.50	12.7000	12.7254	12.7508	12.7762	12.8016	12.8270	12.8524	12.8778	12.9032	12.9286
.51	12.9540	12.9794	13.0048	13.0302	13.0556	13.0810	13.1064	13.1318	13.1572	13.1826
.52	13.2080	13.2334	13.2588	13.2842	13.3096	13.3350	13.3604	13.3858	13.4112	13.4366
.53	13.4620	13.4874	13.5128	13.5382	13.5636	13.5890	13.6144	13.6398	13.6652	13.6906
.54	13.7160	13.7414	13.7668	13.7922	13.8176	13.8430	13.8684	13.8938	13.9192	13.9446
.55	13.9700	13.9954	14.0208	14.0462	14.0716	14.0970	14.1224	14.1478	14.1732	14.1986
.56	14.2240	14.2494	14.2748	14.3002	14.3256	14.3510	14.3764	14.4018	14.4272	14.4526
.57	14.4780	14.5034	14.5288	14.5542	14.5796	14.6050	14.6304	14.6558	14.6812	14.7066
.58	14.7320	14.7574	14.7828	14.8082	14.8336	14.8590	14.8844	14.9098	14.9352	14.9606
.59	14.9860	15.0114	15.0368	15.0622	15.0876	15.1130	15.1384	15.1638	15.1892	15.2146
.60	15.2400	15.2654	15.2908	15.3162	15.3416	15.3670	15.3924	15.4178	15.4432	15.4686
.61	15.4940	15.5194	15.5448	15.5702	15.5956	15.6210	15.6464	15.6718	15.6972	15.7226
.62	15.7480	15.7734	15.7988	15.8242	15.8496	15.8750	15.9004	15.9258	15.9512	15.9766
.63	16.0020	16.0274	16.0528	16.0782	16.1036	16.1290	16.1544	16.1798	16.2052	16.2306
.64	16.2560	16.2814	16.3068	16.3322	16.3576	16.3830	16.4084	16.4338	16.4592	16.4846
.65	16.5100	16.5354	16.5608	16.5862	16.6116	16.6370	16.6624	16.6878	16.7132	16.7386
.66	16.7640	16.7894	16.8148	16.8402	16.8656	16.8910	16.9164	16.9418	16.9672	16.9926
.67	17.0180	17.0434	17.0688	17.0942	17.1196	17.1450	17.1704	17.1958	17.2212	17.2466
.68	17.2720	17.2974	17.3228	17.3482	17.3736	17.3990	17.4244	17.4498	17.4752	17.5006
.69	17.5260	17.5514	17.5768	17.6022	17.6276	17.6530	17.6784	17.7038	17.7292	17.7546
.70	17.7800	17.8054	17.8308	17.8562	17.8816	17.9070	17.9324	17.9578	17.9832	18.0086
.71	18.0340	18.0594	18.0848	18.1102	18.1356	18.1610	18.1864	18.2118	18.2372	18.2626
.72	18.2880	18.3134	18.3388	18.3642	18.3896	18.4150	18.4404	18.4658	18.4912	18.5166
.73	18.5420	18.5674	18.5928	18.6182	18.6436	18.6690	18.6944	18.7198	18.7452	18.7706
.74	18.7960	18.8214	18.8468	18.8722	18.8976	18.9230	18.9484	18.9738	18.9992	19.0246
.75	19.0500	19.0754	19.1008	19.1262	19.1516	19.1770	19.2024	19.2278	19.2532	19.2786
.76	19.3040	19.3294	19.3548	19.3802	19.4056	19.4310	19.4564	19.4818	19.5072	19.5326
.77	19.5580	19.5834	19.6088	19.6342	19.6596	19.6850	19.7104	19.7358	19.7612	19.7866
.78	19.8120	19.8374	19.8628	19.8882	19.9136	19.9390	19.9644	19.9898	20.0152	20.0406
.79	20.0660	20.0914	20.1168	20.1422	20.1676	20.1930	20.2184	20.2438	20.2692	20.2946
.80	20.3200	20.3454	20.3708	20.3962	20.4216	20.4470	20.4724	20.4978	20.5232	20.5486
.81	20.5740	20.5994	20.6248	20.6502	20.6756	20.7010	20.7264	20.7518	20.7772	20.8026
.82	20.8280	20.8534	20.8788	20.9042	20.9296	20.9550	20.9804	21.0058	21.0312	21.0566
.83	21.0820	21.1074	21.1328	21.1582	21.1836	21.2090	21.2344	21.2598	21.2852	21.3106
.84	21.3360	21.3614	21.3868	21.4122	21.4376	21.4630	21.4884	21.5138	21.5392	21.5646
.85	21.5900	21.6154	21.6408	21.6662	21.6916	21.7170	21.7424	21.7678	21.7932	21.8186
.86	21.8440	21.8694	21.8948	21.9202	21.9456	21.9710	21.9964	22.0218	22.0472	22.0726
.87	22.0980	22.1234	22.1488	22.1742	22.1996	22.2250	22.2504	22.2758	22.3012	22.3266
.88	22.3520	22.3774	22.4028	22.4282	22.4536	22.4790	22.5044	22.5298	22.5552	22.5806
.89	22.6060	22.6314	22.6568	22.6822	22.7076	22.7330	22.7584	22.7838	22.8092	22.8346
.90	22.8600	22.8854	22.9108	22.9362	22.9616	22.9870	23.0124	23.0378	23.0632	23.0886
.91	23.1140	23.1394	23.1648	23.1902	23.2156	23.2410	23.2664	23.2918	23.3172	23.3426
.92	23.3680	23.3934	23.4188	23.4442	23.4696	23.4950	23.5204	23.5458	23.5712	23.5966
.93	23.6220	23.6474	23.6728	23.6982	23.7236	23.7490	23.7744	23.7998	23.8252	23.8506
.94	23.8760	23.9014	23.9268	23.9522	23.9776	24.0030	24.0284	24.0538	24.0792	24.1046
.95	24.1300	24.1554	24.1808	24.2062	24.2316	24.2570	24.2824	24.3078	24.3332	24.3586
.96	24.3840	24.4094	24.4348	24.4602	24.4856	24.5110	24.5364	24.5618	24.5872	24.6126
.97	24.6380	24.6634	24.6888	24.7142	24.7396	24.7650	24.7904	24.8158	24.8412	24.8666
.98	24.8920	24.9174	24.9428	24.9682	24.9936	25.0190	25.0444	25.0698	25.0952	25.1206
.99	25.1460	25.1714	25.1968	25.2222	25.2476	25.2730	25.2984	25.3238	25.3492	25.3746
1.00	25.4000	25.4254	25.4508	25.4762	25.5016	25.5270	25.5524	25.5778	25.6032	25.6286

INCREMENTS OF .0001 INCH   .0001 TO .0009 INCH

.0001	.0002	.0003	.0004	.0005	.0006	.0007	.0008	.0009
.00254	.00508	.00762	.01016	.01270	.01524	.01778	.02032	.02286

(Caterpillar)

APPENDIX

MILLIMETERS TO INCHES

INCREMENTS OF 1000 MILLIMETERS 1000-9000 MILLIMETERS

	1000	2000	3000	4000	5000	6000	7000	8000	9000
	39.37008	78.74016	118.11024	157.48031	196.85039	236.22047	275.59055	314.96063	354.33071

INCREMENTS OF 10 MILLIMETERS 0-1090 MILLIMETERS

	0	10	20	30	40	50	60	70	80	90
0	0	.39370	.78740	1.18110	1.57480	1.96850	2.36220	2.75591	3.14961	3.54331
100	3.93701	4.33071	4.72441	5.11811	5.51181	5.90551	6.29921	6.69291	7.08661	7.48031
200	7.87402	8.26772	8.66142	9.05512	9.44882	9.84252	10.23622	10.62992	11.02362	11.41732
300	11.81102	12.20472	12.59843	12.99213	13.38583	13.77953	14.17323	14.56693	14.96063	15.35433
400	15.74803	16.14173	16.53543	16.92913	17.32283	17.71654	18.11024	18.50394	18.89764	19.29134
500	19.68504	20.07874	20.47244	20.86614	21.25984	21.65354	22.04724	22.44094	22.83465	23.22835
600	23.62205	24.01575	24.40945	24.80315	25.19685	25.59055	25.98425	26.37795	26.77165	27.16535
700	27.55906	27.95276	28.34646	28.74016	29.13386	29.52756	29.92126	30.31496	30.70866	31.10236
800	31.49606	31.88976	32.28346	32.67717	33.07087	33.46457	33.85827	34.25197	34.64567	35.03937
900	35.43307	35.82677	36.22047	36.61417	37.00787	37.40157	37.79528	38.18898	38.58268	38.97638
1000	39.37008	39.76378	40.15748	40.55118	40.94488	41.33858	41.73228	42.12598	42.51969	42.91339

INCREMENTS OF .01 MILLIMETERS 0-10.09 MILLIMETERS

	.00	.01	.02	.03	.04	.05	.06	.07	.08	.09
0	0	.00039	.00079	.00118	.00157	.00197	.00236	.00276	.00315	.00354
.1	.00394	.00433	.00472	.00512	.00551	.00591	.00630	.00669	.00709	.00748
.2	.00787	.00827	.00866	.00906	.00945	.00984	.01024	.01063	.01102	.01142
.3	.01181	.01220	.01260	.01299	.01339	.01378	.01417	.01457	.01496	.01535
.4	.01575	.01614	.01654	.01693	.01732	.01772	.01811	.01850	.01890	.01929
.5	.01969	.02008	.02047	.02087	.02126	.02165	.02205	.02244	.02283	.02323
.6	.02362	.02402	.02441	.02480	.02520	.02559	.02598	.02638	.02677	.02717
.7	.02756	.02795	.02835	.02874	.02913	.02953	.02992	.03031	.03071	.03110
.8	.03150	.03189	.03228	.03268	.03307	.03346	.03386	.03425	.03465	.03504
.9	.03543	.03583	.03622	.03661	.03701	.03740	.03780	.03819	.03858	.03898
1.0	.03937	.03976	.04016	.04055	.04094	.04134	.04173	.04213	.04252	.04291
1.1	.04331	.04370	.04409	.04449	.04488	.04528	.04567	.04606	.04646	.04685
1.2	.04724	.04764	.04803	.04843	.04882	.04921	.04961	.05000	.05039	.05079
1.3	.05118	.05157	.05197	.05236	.05276	.05315	.05354	.05394	.05433	.05472
1.4	.05512	.05551	.05591	.05630	.05669	.05709	.05748	.05787	.05827	.05866
1.5	.05906	.05945	.05984	.06024	.06063	.06102	.06142	.06181	.06220	.06260
1.6	.06299	.06339	.06378	.06417	.06457	.06496	.06535	.06575	.06614	.06654
1.7	.06693	.06732	.06772	.06811	.06850	.06890	.06929	.06969	.07008	.07047
1.8	.07087	.07126	.07165	.07205	.07244	.07283	.07323	.07362	.07402	.07441
1.9	.07480	.07520	.07559	.07598	.07638	.07677	.07717	.07756	.07795	.07835
2.0	.07874	.07913	.07953	.07992	.08031	.08071	.08110	.08150	.08189	.08228
2.1	.08268	.08307	.08346	.08386	.08425	.08465	.08504	.08543	.08583	.08622
2.2	.08661	.08701	.08740	.08780	.08819	.08858	.08898	.08937	.08976	.09016
2.3	.09055	.09094	.09134	.09173	.09213	.09252	.09291	.09331	.09370	.09409
2.4	.09449	.09488	.09528	.09567	.09606	.09646	.09685	.09724	.09764	.09803
2.5	.09843	.09882	.09921	.09961	.10000	.10039	.10079	.10118	.10157	.10197
2.6	.10236	.10276	.10315	.10354	.10394	.10433	.10472	.10512	.10551	.10591
2.7	.10630	.10669	.10709	.10748	.10787	.10827	.10866	.10906	.10945	.10984
2.8	.11024	.11063	.11102	.11142	.11181	.11220	.11260	.11299	.11339	.11378
2.9	.11417	.11457	.11496	.11535	.11575	.11614	.11654	.11693	.11732	.11772
3.0	.11811	.11850	.11890	.11929	.11969	.12008	.12047	.12087	.12126	.12165
3.1	.12205	.12244	.12283	.12323	.12362	.12402	.12441	.12480	.12520	.12559
3.2	.12598	.12638	.12677	.12717	.12756	.12795	.12835	.12874	.12913	.12953
3.3	.12992	.13031	.13071	.13110	.13150	.13189	.13228	.13268	.13307	.13346
3.4	.13386	.13425	.13465	.13504	.13543	.13583	.13622	.13661	.13701	.13740
3.5	.13780	.13819	.13858	.13898	.13937	.13976	.14016	.14055	.14094	.14134
3.6	.14173	.14213	.14252	.14291	.14331	.14370	.14409	.14449	.14488	.14528
3.7	.14567	.14606	.14646	.14685	.14724	.14764	.14803	.14843	.14882	.14921
3.8	.14961	.15000	.15039	.15079	.15118	.15157	.15197	.15236	.15276	.15315
3.9	.15354	.15394	.15433	.15472	.15512	.15551	.15591	.15630	.15669	.15709
4.0	.15748	.15787	.15827	.15866	.15906	.15945	.15984	.16024	.16063	.16102
4.1	.16142	.16181	.16220	.16260	.16299	.16339	.16378	.16417	.16457	.16496
4.2	.16535	.16575	.16614	.16654	.16693	.16732	.16772	.16811	.16850	.16890
4.3	.16929	.16969	.17008	.17047	.17087	.17126	.17165	.17205	.17244	.17283
4.4	.17323	.17362	.17402	.17441	.17480	.17520	.17559	.17598	.17638	.17677
4.5	.17717	.17756	.17795	.17835	.17874	.17913	.17953	.17992	.18031	.18071
4.6	.18110	.18150	.18189	.18228	.18268	.18307	.18346	.18386	.18425	.18465
4.7	.18504	.18543	.18583	.18622	.18661	.18701	.18740	.18780	.18819	.18858
4.8	.18898	.18937	.18976	.19016	.19055	.19094	.19134	.19173	.19213	.19252
4.9	.19291	.19331	.19370	.19409	.19449	.19488	.19528	.19567	.19606	.19646
5.0	.19685	.19724	.19764	.19803	.19843	.19882	.19921	.19961	.20000	.20039
5.1	.20079	.20118	.20157	.20197	.20236	.20276	.20315	.20354	.20394	.20433
5.2	.20472	.20512	.20551	.20591	.20630	.20669	.20709	.20748	.20787	.20827
5.3	.20866	.20906	.20945	.20984	.21024	.21063	.21102	.21142	.21181	.21220
5.4	.21260	.21299	.21339	.21378	.21417	.21457	.21496	.21535	.21575	.21614
5.5	.21654	.21693	.21732	.21772	.21811	.21850	.21890	.21929	.21969	.22008
5.6	.22047	.22087	.22126	.22165	.22205	.22244	.22283	.22323	.22362	.22402
5.7	.22441	.22480	.22520	.22559	.22598	.22638	.22677	.22717	.22756	.22795
5.8	.22835	.22874	.22913	.22953	.22992	.23031	.23071	.23110	.23150	.23189
5.9	.23228	.23268	.23307	.23346	.23386	.23425	.23465	.23504	.23543	.23583
6.0	.23622	.23661	.23701	.23740	.23780	.23819	.23858	.23898	.23937	.23976
6.1	.24016	.24055	.24094	.24134	.24173	.24213	.24252	.24291	.24331	.24370
6.2	.24409	.24449	.24488	.24528	.24567	.24606	.24646	.24685	.24724	.24764
6.3	.24803	.24843	.24882	.24921	.24961	.25000	.25039	.25079	.25118	.25157
6.4	.25197	.25236	.25276	.25315	.25354	.25394	.25433	.25472	.25512	.25551
6.5	.25591	.25630	.25669	.25709	.25748	.25787	.25827	.25866	.25906	.25945
6.6	.25984	.26024	.26063	.26102	.26142	.26181	.26220	.26260	.26299	.26339
6.7	.26378	.26417	.26457	.26496	.26535	.26575	.26614	.26654	.26693	.26732
6.8	.26772	.26811	.26850	.26890	.26929	.26969	.27008	.27047	.27087	.27126
6.9	.27165	.27205	.27244	.27283	.27323	.27362	.27402	.27441	.27480	.27520
7.0	.27559	.27598	.27638	.27677	.27717	.27756	.27795	.27835	.27874	.27913
7.1	.27953	.27992	.28031	.28071	.28110	.28150	.28189	.28228	.28268	.28307
7.2	.28346	.28386	.28425	.28465	.28504	.28543	.28583	.28622	.28661	.28701
7.3	.28740	.28780	.28819	.28858	.28898	.28937	.28976	.29016	.29055	.29094
7.4	.29134	.29173	.29213	.29252	.29291	.29331	.29370	.29409	.29449	.29488
7.5	.29528	.29567	.29606	.29646	.29685	.29724	.29764	.29803	.29843	.29882
7.6	.29921	.29961	.30000	.30039	.30079	.30118	.30157	.30197	.30236	.30276
7.7	.30315	.30354	.30394	.30433	.30472	.30512	.30551	.30591	.30630	.30669
7.8	.30709	.30748	.30787	.30827	.30866	.30906	.30945	.30984	.31024	.31063
7.9	.31102	.31142	.31181	.31220	.31260	.31299	.31339	.31378	.31417	.31457
8.0	.31496	.31535	.31575	.31614	.31654	.31693	.31732	.31772	.31811	.31850
8.1	.31890	.31929	.31969	.32008	.32047	.32087	.32126	.32165	.32205	.32244
8.2	.32283	.32323	.32362	.32402	.32441	.32480	.32520	.32559	.32598	.32638
8.3	.32677	.32717	.32756	.32795	.32835	.32874	.32913	.32953	.32992	.33031
8.4	.33071	.33110	.33150	.33189	.33228	.33268	.33307	.33346	.33386	.33425
8.5	.33465	.33504	.33543	.33583	.33622	.33661	.33701	.33740	.33780	.33819
8.6	.33858	.33898	.33937	.33976	.34016	.34055	.34094	.34134	.34173	.34213
8.7	.34252	.34291	.34331	.34370	.34409	.34449	.34488	.34528	.34567	.34606
8.8	.34646	.34685	.34724	.34764	.34803	.34843	.34882	.34921	.34961	.35000
8.9	.35039	.35079	.35118	.35157	.35197	.35236	.35276	.35315	.35354	.35394
9.0	.35433	.35472	.35512	.35551	.35591	.35630	.35669	.35709	.35748	.35787
9.1	.35827	.35866	.35906	.35945	.35984	.36024	.36063	.36102	.36142	.36181
9.2	.36220	.36260	.36299	.36339	.36378	.36417	.36457	.36496	.36535	.36575
9.3	.36614	.36654	.36693	.36732	.36772	.36811	.36850	.36890	.36929	.36969
9.4	.37008	.37047	.37087	.37126	.37165	.37205	.37244	.37283	.37323	.37362
9.5	.37402	.37441	.37480	.37520	.37559	.37598	.37638	.37677	.37717	.37756
9.6	.37795	.37835	.37874	.37913	.37953	.37992	.38031	.38071	.38110	.38150
9.7	.38189	.38228	.38268	.38307	.38346	.38386	.38425	.38465	.38504	.38543
9.8	.38583	.38622	.38661	.38701	.38740	.38780	.38819	.38858	.38898	.38937
9.9	.38976	.39016	.39055	.39094	.39134	.39173	.39213	.39252	.39291	.39331
10.0	.39370	.39409	.39449	.39488	.39528	.39567	.39606	.39646	.39685	.39724

INCREMENTS OF .001 MILLIMETERS .001-.009 MILLIMETERS

	.001	.002	.003	.004	.005	.006	.007	.008	.009
	.00004	.00008	.00012	.00016	.00020	.00024	.00028	.00031	.00035

(Caterpillar)

AREA
Square Measure Conversion*

in.²	cm²	cm²	in.²	ft.²	m²	m²	ft.²	yd.²	m²	m²	yd.²	acres	hectares	hectares	acres
1	6.45	1	0.16	1	0.09	1	10.76	1	0.84	1	1.20	1	0.40	1	2.47
2	12.90	2	0.31	2	0.19	2	21.53	2	1.67	2	2.39	2	0.81	2	4.94
3	19.35	3	0.47	3	0.27	3	32.29	3	2.51	3	3.59	3	1.21	3	7.41
4	25.81	4	0.62	4	0.37	4	43.06	4	3.34	4	4.78	4	1.62	4	9.88
5	32.26	5	0.78	5	0.46	5	53.82	5	4.18	5	5.98	5	2.02	5	12.36
6	38.71	6	0.93	6	0.56	6	64.58	6	5.02	6	7.18	6	2.43	6	14.83
7	45.16	7	1.09	7	0.65	7	75.35	7	5.85	7	8.37	7	2.83	7	17.30
8	51.61	8	1.24	8	0.74	8	86.11	8	6.69	8	9.57	8	3.24	8	19.77
9	58.06	9	1.40	9	0.84	9	96.88	9	7.53	9	10.76	9	3.64	9	22.24
10	64.52	10	1.55	10	0.93	10	107.64	10	8.36	10	11.56	10	4.05	10	24.71
11	70.97	11	1.71	11	1.02	11	118.40	11	9.20	11	13.16	11	4.45	11	27.18
12	77.42	12	1.86	12	1.11	12	129.17	12	10.03	12	14.35	12	4.86	12	29.65
13	83.87	13	2.02	13	1.21	13	139.93	13	10.87	13	15.55	13	5.26	13	32.12
14	90.32	14	2.17	14	1.30	14	150.70	14	11.71	14	16.74	14	5.67	14	34.59
15	96.77	15	2.33	15	1.39	15	161.46	15	12.54	15	17.94	15	6.07	15	37.07
16	103.22	16	2.48	16	1.49	16	172.22	16	13.38	16	19.14	16	6.47	16	39.54
17	109.68	17	2.64	17	1.58	17	182.99	17	14.21	17	20.33	17	6.88	17	42.01
18	116.13	18	2.79	18	1.67	18	193.75	18	15.05	18	21.53	18	7.28	18	44.48
19	122.58	19	2.95	19	1.77	19	204.51	19	15.89	19	22.72	19	7.69	19	46.95
20	129.03	20	3.10	20	1.86	20	215.28	20	16.72	20	23.92	20	8.09	20	49.42
21	135.48	21	3.26	21	1.95	21	226.04	21	17.56	21	25.12	21	8.50	21	51.89
22	141.93	22	3.41	22	2.04	22	236.81	22	18.39	22	26.31	22	8.90	22	54.36
23	148.39	23	3.57	23	2.14	23	247.57	23	19.23	23	27.51	23	9.31	23	56.83
24	154.84	24	3.72	24	2.23	24	258.33	24	20.07	24	28.70	24	9.71	24	59.31
25	161.29	25	3.88	25	2.32	25	269.10	25	20.90	25	29.90	25	10.12	25	61.78
26	167.74	26	4.03	26	2.42	26	279.86	26	21.74	26	31.10	26	10.52	26	64.25
27	174.19	27	4.19	27	2.51	27	290.63	27	22.58	27	32.29	27	10.93	27	66.72
28	180.64	28	4.34	28	2.60	28	301.39	28	23.41	28	33.49	28	11.33	28	69.19
29	187.10	29	4.50	29	2.69	29	312.15	29	24.25	29	34.68	29	11.74	29	71.66
30	193.55	30	4.65	30	2.79	30	322.92	30	25.08	30	35.88	30	12.14	30	74.13
31	200.00	31	4.81	31	2.88	31	333.68	31	25.92	31	37.08	31	12.55	31	76.60
32	206.45	32	4.96	32	2.97	32	344.45	32	26.76	32	38.27	32	12.95	32	79.07
33	212.90	33	5.12	33	3.07	33	355.21	33	27.59	33	39.47	33	13.35	33	81.54
34	219.35	34	5.27	34	3.16	34	365.97	34	28.43	34	40.66	34	13.76	34	84.01
35	225.81	35	5.43	35	3.25	35	376.74	35	29.26	35	41.86	35	14.16	35	86.49
36	232.26	36	5.58	36	3.34	36	387.50	36	30.10	36	43.06	36	14.57	36	88.96
37	238.71	37	5.74	37	3.44	37	398.27	37	30.94	37	44.25	37	14.97	37	91.43
38	245.16	38	5.89	38	3.53	38	409.03	38	31.77	38	45.44	38	15.38	38	93.90
39	251.61	39	6.05	39	3.62	39	419.79	39	32.61	39	46.64	39	15.78	39	96.37
40	258.06	40	6.20	40	3.72	40	430.56	40	33.45	40	47.84	40	16.19	40	98.84
41	264.52	41	6.36	41	3.81	41	441.32	41	34.28	41	49.04	41	16.59	41	101.31
42	270.97	42	6.51	42	3.90	42	452.08	42	35.12	42	50.23	42	17.00	42	103.78
43	277.42	43	6.67	43	3.99	43	462.85	43	35.95	43	51.43	43	17.40	43	106.26
44	283.87	44	6.82	44	4.09	44	473.61	44	36.79	44	52.62	44	17.81	44	108.73
45	290.32	45	6.98	45	4.18	45	484.38	45	37.63	45	53.82	45	18.21	45	111.10
46	296.77	46	7.13	46	4.27	46	495.14	46	38.46	46	55.02	46	18.62	46	113.67
47	303.23	47	7.29	47	4.37	47	505.90	47	39.30	47	56.21	47	19.02	47	116.14
48	309.68	48	7.44	48	4.46	48	516.67	48	40.13	48	57.41	48	19.42	48	118.61
49	316.13	49	7.60	49	4.55	49	527.43	49	40.97	49	58.60	49	19.83	49	121.08

(CONTINUED)

AREA (continued)
Square Measure Conversion

No.	A	No.	B	No.	C	No.	D	No.	E	No.	F	No.	G	No.	H
50	322.58	50	7.75	50	4.65	50	538.20	50	41.81	50	59.80	50	20.23	50	123.55
51	329.03	51	7.91	51	4.74	51	548.96	51	42.64	51	61.00	51	20.64	51	126.02
52	335.48	52	8.06	52	4.83	52	559.72	52	43.48	52	62.19	52	21.04	52	128.49
53	341.93	53	8.22	53	4.92	53	570.49	53	44.31	53	63.39	53	21.45	53	130.97
54	348.39	54	8.37	54	5.02	54	581.25	54	45.15	54	64.58	54	21.85	54	133.44
55	354.84	55	8.53	55	5.11	55	592.02	55	45.99	55	65.78	55	22.26	55	135.91
56	361.29	56	8.68	56	5.20	56	602.78	56	46.82	56	66.98	56	22.66	56	138.38
57	367.74	57	8.84	57	5.30	57	613.54	57	47.66	57	68.17	57	23.07	57	140.85
58	374.19	58	8.99	58	5.39	58	624.31	58	48.50	58	69.37	58	23.47	58	143.32
59	380.64	59	9.15	59	5.48	59	635.07	59	49.33	59	70.56	59	23.88	59	145.79
60	387.10	60	9.30	60	5.57	60	645.84	60	50.17	60	71.76	60	24.28	60	148.26
61	393.55	61	9.46	61	5.67	61	656.60	61	51.00	61	72.96	61	24.69	61	150.73
62	400.00	62	9.61	62	5.76	62	667.36	62	51.84	62	74.15	62	25.09	62	153.21
63	406.45	63	9.77	63	5.85	63	678.13	63	52.68	63	75.35	63	25.50	63	155.68
64	412.90	64	9.92	64	5.95	64	688.89	64	53.51	64	76.54	64	25.90	64	158.15
65	419.35	65	10.08	65	6.04	65	699.65	65	54.35	65	77.74	65	26.30	65	160.62
66	325.81	66	10.23	66	6.13	66	710.42	66	55.18	66	78.94	66	26.71	66	163.09
67	432.26	67	10.39	67	6.22	67	721.18	67	56.02	67	80.13	67	27.11	67	165.56
68	438.71	68	10.54	68	6.32	68	731.95	68	56.86	68	81.33	68	27.52	68	168.03
69	445.16	69	10.70	69	6.41	69	742.71	69	57.69	69	82.52	69	27.92	69	170.50
70	451.61	70	10.85	70	6.50	70	753.47	70	58.53	70	83.72	70	28.33	70	172.97
71	458.06	71	11.01	71	6.60	71	764.24	71	59.37	71	84.92	71	28.73	71	175.44
72	464.51	72	11.16	72	6.69	72	775.00	72	60.20	72	86.11	72	29.14	72	177.92
73	470.97	73	11.32	73	6.78	73	785.77	73	61.04	73	87.31	73	29.54	73	180.39
74	477.42	74	11.47	74	6.87	74	796.53	74	61.87	74	88.50	74	29.95	74	182.86
75	483.87	75	11.63	75	6.97	75	807.29	75	62.70	75	89.70	75	30.35	75	185.33
76	490.32	76	11.78	76	7.06	76	818.06	76	63.55	76	90.90	76	30.76	76	187.80
77	496.77	77	11.94	77	7.15	77	828.82	77	64.38	77	92.09	77	31.16	77	190.27
78	503.22	78	12.09	78	7.25	78	839.59	78	65.22	78	93.29	78	31.57	78	192.74
79	509.68	79	12.25	79	7.34	79	850.35	79	66.05	79	94.48	79	31.97	79	195.21
80	516.13	80	12.40	80	7.43	80	861.11	80	66.89	80	95.68	80	32.37	80	197.68
81	522.58	81	12.56	81	7.53	81	871.88	81	67.73	81	96.88	81	32.78	81	200.16
82	529.03	82	12.71	82	7.62	82	882.64	82	68.56	82	98.07	82	33.18	82	202.63
83	535.48	83	12.87	83	7.71	83	893.41	83	69.40	83	99.27	83	33.59	83	205.10
84	541.93	84	13.02	84	7.80	84	904.17	84	70.23	84	100.46	84	33.99	84	207.57
85	548.39	85	13.18	85	7.90	85	914.93	85	71.07	85	101.66	85	34.40	85	210.04
86	554.84	86	13.33	86	7.99	86	925.70	86	71.91	86	102.86	86	34.80	86	212.51
87	561.29	87	13.49	87	8.08	87	936.46	87	72.74	87	104.05	87	35.21	87	214.98
88	567.74	88	13.64	88	8.17	88	947.22	88	73.58	88	105.25	88	35.61	88	217.45
89	574.19	89	13.80	89	8.27	89	957.99	89	74.42	89	106.44	89	36.02	89	219.92
90	580.64	90	13.95	90	8.36	90	968.75	90	75.25	90	107.64	90	36.42	90	222.39
91	587.10	91	14.11	91	8.45	91	979.52	91	76.09	91	108.84	91	36.83	91	224.87
92	593.55	92	14.26	92	8.55	92	990.28	92	76.92	92	110.03	92	37.23	92	227.34
93	600.00	93	14.42	93	8.64	93	1001.04	93	77.76	93	111.23	93	37.64	93	229.81
94	606.45	94	14.57	94	8.73	94	1011.81	94	78.60	94	112.42	94	38.04	94	232.28
95	612.90	95	14.73	95	8.83	95	1022.57	95	79.43	95	113.62	95	38.45	95	234.75
96	619.35	96	14.88	96	8.92	96	1033.34	96	80.27	96	114.82	96	38.84	96	237.22
97	625.80	97	15.04	97	9.01	97	1044.10	97	81.10	97	116.01	97	39.25	97	239.69
98	632.26	98	15.19	98	9.10	98	1054.86	98	81.94	98	117.21	98	39.66	98	242.16
99	638.71	99	15.35	99	9.20	99	1065.63	99	82.78	99	118.40	99	40.06	99	244.63
100	645.16	100	15.50	100	9.29	100	1076.39	100	83.61	100	119.60	100	40.47	100	247.11

*Adapted from Units of Weight and Measure, National Bureau of Standards.

MASS
Avoirdupois (Dry) Ounces and Pounds to Kilograms*

Ounces	1	2	3	4	5	6	7	8	9	10	11	12	13	14	15
Kilograms	.028	.057	.085	.113	.142	.170	.198	.227	.255	.283	.312	.340	.369	.397	.425

AVDP POUNDS	KILO-GRAMS	AVDP POUNDS	KILO-GRAMS	AVDP POUNDS	KILO-GRAMS	AVDP POUNDS	KILO-GRAMS	AVDP POUNDS	KILO-GRAMS
0	0.000	100	45.359	200	90.718	300	136.078	400	181.437
1	0.454	101	45.813	201	91.172	301	136.531	401	181.891
2	0.907	102	46.266	202	91.626	302	136.985	402	182.344
3	1.361	103	46.720	203	92.079	303	137.438	403	182.798
4	1.814	104	47.174	204	92.533	304	137.892	404	183.251
5	2.268	105	47.627	205	92.986	305	138.346	405	183.705
6	2.722	106	48.081	206	93.440	306	138.799	406	184.159
7	3.175	107	48.534	207	93.894	307	139.253	407	184.612
8	3.629	108	48.988	208	94.347	308	139.706	408	185.066
9	4.082	109	49.442	209	94.801	309	140.160	409	185.519
10	4.536	110	49.895	210	95.254	310	140.614	410	185.973
11	4.990	111	50.349	211	95.708	311	141.067	411	186.426
12	5.443	112	50.802	212	96.162	312	141.521	412	186.880
13	5.897	113	51.256	213	96.615	313	141.974	413	187.334
14	6.350	114	51.710	214	97.069	314	142.428	414	187.787
15	6.804	115	52.163	215	97.522	315	142.882	415	188.241
16	7.257	116	52.617	216	97.976	316	143.335	416	188.694
17	7.711	117	53.070	217	98.430	317	143.789	417	189.148
18	8.165	118	53.524	218	98.883	318	144.242	418	189.602
19	8.618	119	53.977	219	99.337	319	144.696	419	190.055
20	9.072	120	54.431	220	99.790	320	145.150	420	190.509
21	9.525	121	54.885	221	100.244	321	145.603	421	190.962
22	9.979	122	55.338	222	100.698	322	146.057	422	191.416
23	10.433	123	55.792	223	101.151	323	146.510	423	191.870
24	10.886	124	56.245	224	101.605	324	146.964	424	192.323
25	11.340	125	56.699	225	102.058	325	147.418	425	192.777
26	11.793	126	57.153	226	102.512	326	147.871	426	193.230
27	12.247	127	57.606	227	102.965	327	148.325	427	193.684
28	12.701	128	58.060	228	103.419	328	148.778	428	194.138
29	13.154	129	58.513	229	103.873	329	149.232	429	194.591
30	13.608	130	58.967	230	104.326	330	149.685	430	195.045
31	14.061	131	59.421	231	104.780	331	150.139	431	195.498
32	14.515	132	59.874	232	105.233	332	150.593	432	195.952
33	14.969	133	60.328	233	105.687	333	151.046	433	196.406
34	15.422	134	60.781	234	106.141	334	151.500	434	196.859
35	15.876	135	61.235	235	106.594	335	151.953	435	197.313
36	16.329	136	61.689	236	107.048	336	152.407	436	197.766
37	16.783	137	62.142	237	107.501	337	152.861	437	198.220
38	17.237	138	62.596	238	107.955	338	153.314	438	198.673
39	17.690	139	63.049	239	108.409	339	153.768	439	199.127
40	18.144	140	63.503	240	108.862	340	154.221	440	199.581
41	18.597	141	63.957	241	109.316	341	154.675	441	200.034
42	19.051	142	64.410	242	109.769	342	155.129	442	200.488
43	19.504	143	64.864	243	110.223	343	155.582	443	200.941
44	19.958	144	65.317	244	110.677	344	156.036	444	201.395
45	20.412	145	65.771	245	111.130	345	156.489	445	201.849
46	20.865	146	66.224	246	111.584	346	156.943	446	202.302
47	21.319	147	66.678	247	112.037	347	157.397	447	202.756
48	21.772	148	67.132	248	112.491	348	157.850	448	203.209
49	22.226	149	67.585	249	112.945	349	158.304	449	203.663

(CONTINUED)

122

MASS (continued)
Avoirdupois (Dry) Ounces and Pounds to Kilograms*

AVDP POUNDS	KILOGRAMS	AVDP POUNDS	KILOGRAMS	AVDP POUNDS	KILOGRAMS	AVDP POUNDS	KILOGRAMS	AVDP POUNDS	KILOGRAMS
50	22.680	150	68.039	250	113.398	350	158.757	450	204.117
51	23.133	151	68.492	251	113.852	351	159.211	451	204.570
52	23.587	152	68.946	252	114.305	352	159.665	452	205.024
53	24.040	153	69.400	253	114.759	353	160.118	453	205.477
54	24.494	154	69.853	254	115.212	354	160.572	454	205.931
55	24.948	155	70.307	255	115.666	355	161.025	455	206.385
56	25.401	156	70.760	256	116.120	356	161.479	456	206.838
57	25.855	157	71.214	257	116.573	357	161.932	457	207.292
58	26.309	158	71.668	258	117.027	358	162.386	458	207.745
59	26.762	159	72.121	259	117.480	359	162.840	459	208.199
60	27.216	160	72.575	260	117.934	360	163.293	460	208.652
61	27.669	161	73.028	261	118.388	361	163.747	461	209.106
62	28.123	162	73.482	262	118.841	362	164.200	462	209.560
63	28.576	163	73.936	263	119.295	363	164.654	463	210.013
64	29.030	164	74.389	264	119.748	364	165.108	464	210.467
65	29.484	165	74.843	265	120.202	365	165.561	465	210.920
66	29.937	166	75.296	266	120.656	366	166.015	466	211.374
67	30.391	167	75.750	267	121.109	367	166.468	467	211.828
68	30.844	168	76.204	268	121.563	368	166.922	468	212.281
69	31.298	169	76.657	269	122.016	369	167.376	469	212.735
70	31.751	170	77.111	270	122.470	370	167.829	470	213.188
71	32.205	171	77.564	271	122.924	371	168.283	471	213.642
72	32.659	172	78.018	272	123.377	372	168.736	472	214.096
73	33.112	173	78.471	273	123.831	373	169.190	473	214.549
74	33.566	174	78.925	274	124.284	374	169.644	474	215.003
75	34.019	175	79.379	275	124.738	375	170.097	475	215.456
76	34.473	176	79.832	276	125.191	376	170.551	476	215.910
77	34.927	177	80.286	277	125.645	377	171.004	477	216.364
78	35.380	178	80.739	278	126.099	378	171.458	478	216.817
79	35.834	179	81.193	279	126.552	379	171.912	479	217.271
80	36.287	180	81.647	280	127.006	380	172.365	480	217.724
81	36.741	181	82.100	281	127.459	381	172.819	481	218.178
82	37.195	182	82.554	282	127.913	382	173.272	482	218.632
83	37.648	183	83.007	283	128.367	383	173.726	483	219.085
84	38.102	184	83.461	284	128.820	384	174.179	484	219.539
85	38.555	185	83.915	285	129.274	385	174.633	485	219.992
86	39.009	186	84.368	286	129.727	386	175.087	486	220.446
87	39.463	187	84.822	287	130.181	387	175.540	487	220.899
88	39.916	188	85.275	288	130.635	388	175.994	488	221.353
89	40.370	189	85.729	289	131.088	389	176.447	489	221.807
90	40.823	190	86.183	290	131.542	390	176.901	490	222.260
91	41.277	191	86.636	291	131.995	391	177.355	491	222.714
92	41.731	192	87.090	292	132.449	392	177.808	492	223.167
93	42.184	193	87.543	293	132.903	393	178.262	493	223.621
94	42.638	194	87.997	294	133.356	394	178.715	494	224.075
95	43.091	195	88.451	295	133.810	395	179.169	495	224.528
96	43.545	196	88.904	296	134.263	396	179.623	496	224.982
97	43.998	197	89.358	297	134.717	397	180.076	497	225.435
98	44.452	198	89.811	298	135.171	398	180.530	498	225.889
99	44.906	199	90.265	299	135.624	399	180.983	499	226.343

(CONTINUED)

* Adapted from Units of Weight and Measure, National Bureau of Standards

MASS (continued)
Avoirdupois (Dry) Ounces and Pounds to Kilograms*

Ounces	1	2	3	4	5	6	7	8	9	10	11	12	13	14	15
Kilograms	.028	.057	.085	.113	.142	.170	.198	.227	.255	.283	.312	.340	.369	.397	.425

AVDP POUNDS	KILO-GRAMS	AVDP POUNDS	KILO-GRAMS	AVDP POUNDS	KILO-GRAMS	AVDP POUNDS	KILO-GRAMS	AVDP POUNDS	KILO-GRAMS
500	226.796	600	272.155	700	317.515	800	362.874	900	408.233
501	227.250	601	272.609	701	317.968	801	363.327	901	408.687
502	227.703	602	273.063	702	318.422	802	363.781	902	409.140
503	228.157	603	273.516	703	318.875	803	364.235	903	409.594
504	228.611	604	273.970	704	319.329	804	364.688	904	410.048
505	229.064	605	274.423	705	319.783	805	365.142	905	410.501
506	229.518	606	274.877	706	320.236	806	365.595	906	410.955
507	229.971	607	275.331	707	320.690	807	366.049	907	411.408
508	230.425	608	275.784	708	321.143	808	366.503	908	411.862
509	230.879	609	276.238	709	321.597	809	366.956	909	412.315
510	231.332	610	276.691	710	322.051	810	367.410	910	412.769
511	231.786	611	277.145	711	322.504	811	367.863	911	413.223
512	232.239	612	277.599	712	322.958	812	368.317	912	413.676
513	232.693	613	278.052	713	323.411	813	368.771	913	414.130
514	233.146	614	278.506	714	323.865	814	369.224	914	414.583
515	233.600	615	278.959	715	324.319	815	369.678	915	415.037
516	234.054	616	279.413	716	324.772	816	370.131	916	415.491
517	234.507	617	279.866	717	325.226	817	370.585	917	415.944
518	234.961	618	280.320	718	325.679	818	371.039	918	416.398
519	235.414	619	280.774	719	326.133	819	371.492	919	416.851
520	235.868	620	281.227	720	326.587	820	371.946	920	417.305
521	236.322	621	281.681	721	327.040	821	372.399	921	417.759
522	236.775	622	282.134	722	327.494	822	372.853	922	418.212
523	237.229	623	282.588	723	327.947	823	373.307	923	418.666
524	237.682	624	283.042	724	328.401	824	373.760	924	419.119
525	238.136	625	283.495	725	328.854	825	374.214	925	419.573
526	238.590	626	283.949	726	329.308	826	374.667	926	420.027
527	239.043	627	284.402	727	329.762	827	375.121	927	420.480
528	239.497	628	284.856	728	330.215	828	375.574	928	420.934
529	239.950	629	285.310	729	330.669	829	376.028	929	421.387
530	240.404	630	285.763	730	331.122	830	376.482	930	421.841
531	240.858	631	286.217	731	331.576	831	376.935	931	422.294
532	241.311	632	286.670	732	332.030	832	377.389	932	422.748
533	241.765	633	287.124	733	332.483	833	377.842	933	423.202
534	242.218	634	287.578	734	332.937	834	378.296	934	423.655
535	242.672	635	288.031	735	333.390	835	378.750	935	424.109
536	243.126	636	288.485	736	333.844	836	379.203	936	424.562
537	243.579	637	288.938	737	334.298	837	379.657	937	425.016
538	244.033	638	289.392	738	334.751	838	380.110	938	425.470
539	244.486	639	289.846	739	335.205	839	380.564	939	425.923
540	244.940	640	290.299	740	335.658	840	381.018	940	426.377
541	245.393	641	290.753	741	336.112	841	381.471	941	426.830
542	245.847	642	291.206	742	336.566	842	381.925	942	427.284
543	246.301	643	291.660	743	337.019	843	382.378	943	427.738
544	246.754	644	292.113	744	337.473	844	382.832	944	428.191
545	247.208	645	292.567	745	337.926	845	383.286	945	428.645
546	247.662	646	293.021	746	338.380	846	383.739	946	429.098
547	248.115	647	293.474	747	338.834	847	384.193	947	429.552
548	248.569	648	293.928	748	339.287	848	384.646	948	430.006
549	249.022	649	294.381	749	339.741	849	385.100	949	430.459

(CONTINUED)

MASS (continued)
Avoirdupois (Dry) Ounces and Pounds to Kilograms*

AVDP POUNDS	KILO-GRAMS	AVDP POUNDS	KILO-GRAMS	AVDP POUNDS	KILO-GRAMS	AVDP POUNDS	KILO-GRAMS	AVDP POUNDS	KILO-GRAMS
550	249.476	650	294.835	750	340.194	850	385.554	950	430.913
551	249.929	651	295.289	751	340.648	851	386.007	951	431.366
552	250.383	652	295.742	752	341.101	852	386.461	952	431.820
553	250.837	653	296.196	753	341.555	853	386.914	953	432.274
554	251.290	654	296.649	754	342.009	854	387.368	954	432.727
555	251.744	655	297.103	755	342.462	855	387.821	955	433.181
556	252.197	656	297.557	756	342.916	856	388.275	956	433.634
557	252.651	657	298.010	757	343.369	857	388.729	957	434.088
558	253.105	658	298.464	758	343.823	858	389.182	958	434.541
559	253.558	659	298.917	759	344.277	859	389.636	959	434.995
560	254.012	660	299.371	760	344.730	860	390.089	960	435.449
561	254.465	661	299.825	761	345.184	861	390.543	961	435.902
562	254.919	662	300.278	762	345.637	862	390.997	962	436.356
563	255.373	663	300.732	763	346.091	863	391.450	963	436.809
564	255.826	664	301.185	764	346.545	864	391.904	964	437.263
565	256.280	665	301.639	765	346.998	865	392.357	965	437.717
566	256.733	666	302.093	766	347.452	866	392.811	966	438.170
567	257.187	667	302.546	767	347.905	867	393.265	967	438.624
568	257.640	668	303.000	768	348.359	868	393.718	968	439.077
569	258.094	669	303.453	769	348.813	869	394.172	969	439.531
570	258.548	670	303.907	770	349.266	870	394.625	970	439.985
571	259.001	671	304.360	771	349.720	871	395.079	971	440.438
572	259.455	672	304.814	772	350.173	872	395.533	972	440.892
573	259.908	673	305.268	773	350.627	873	395.986	973	441.345
574	260.362	674	305.721	774	351.080	874	396.440	974	441.799
575	260.816	675	306.175	775	351.534	875	396.893	975	442.253
576	261.269	676	306.628	776	351.988	876	397.347	976	442.706
577	261.723	677	307.082	777	352.441	877	397.801	977	443.160
578	262.176	678	307.536	778	352.895	878	398.254	978	443.613
579	262.630	679	307.989	779	353.348	879	398.708	979	444.067
580	263.084	680	308.443	780	353.802	880	399.161	980	444.521
581	263.537	681	308.896	781	354.256	881	399.615	981	444.974
582	263.991	682	309.350	782	354.709	882	400.068	982	445.428
583	264.444	683	309.804	783	355.163	883	400.522	983	445.881
584	264.898	684	310.257	784	355.616	884	400.976	984	446.335
585	265.352	685	310.711	785	356.070	885	401.429	985	446.788
586	265.805	686	311.164	786	356.524	886	401.883	986	447.242
587	266.259	687	311.618	787	356.977	887	402.336	987	447.696
588	266.712	688	312.072	788	357.431	888	402.790	988	448.149
589	267.166	689	312.525	789	357.884	889	403.244	989	448.603
590	267.620	690	312.979	790	358.338	890	403.697	990	449.056
591	268.073	691	313.432	791	358.792	891	404.151	991	449.510
592	268.527	692	313.886	792	359.245	892	404.604	992	449.964
593	268.980	693	314.340	793	359.699	893	405.058	993	450.417
594	269.434	694	314.793	794	360.152	894	405.512	994	450.871
595	269.887	695	315.247	795	360.606	895	405.965	995	451.324
596	270.341	696	315.700	796	361.060	896	406.419	996	451.778
597	270.795	697	316.154	797	361.513	897	406.872	997	452.232
598	271.248	698	316.607	798	361.967	898	407.326	998	452.685
599	271.702	699	317.061	799	362.420	899	407.780	999	453.139

* Adapted from Units of Weight and Measure, National Bureau of Standards

VOLUME—CAPACITY
LIQUID OUNCES, QUARTS, AND GALLONS TO LITERS*

	1	2	3	4	5	6	7	8	9	10	11	12	13	14	15	16
Ounces	1	2	3	4	5	6	7	8	9	10	11	12	13	14	15	16
Liters	.030	.059	.089	.118	.148	.177	.207	.237	.266	.296	.325	.355	.385	.414	.444	.473
Ounces	17	18	19	20	21	22	23	24	25	26	27	28	29	30	31	32
Liters	.503	.532	.562	.592	.621	.651	.680	.710	.740	.769	.799	.828	.858	.887	.917	.946
Quarts	1	2	3	4	5	6	7	8	9	10	11	12	13	14	15	16
Liters	.946	1.893	2.839	3.785	4.732	5.678	6.624	7.571	8.517	9.464	10.410	11.356	12.303	13.249	14.195	15.142
Quarts	17	18	19	20	21	22	23	24	25	26	27	28	29	30	31	32
Liters	16.088	17.034	17.981	18.927	19.873	20.820	21.766	22.712	23.659	24.605	25.552	26.498	27.444	28.391	29.337	30.283

GALLONS	LITERS	GALLONS	LITERS	GALLONS	LITERS	GALLONS	LITERS	GALLONS	LITERS
0	0.000	100	378.541	200	757.082	300	1135.624	400	1514.165
1	3.785	101	382.327	201	760.868	301	1139.409	401	1517.950
2	7.571	102	386.112	202	764.653	302	1143.194	402	1521.736
3	11.356	103	389.897	203	768.439	303	1146.980	403	1525.521
4	15.142	104	393.683	204	772.224	304	1150.765	404	1529.306
5	18.927	105	397.468	205	776.009	305	1154.551	405	1533.092
6	22.713	106	401.254	206	779.795	306	1158.336	406	1536.877
7	26.498	107	405.039	207	783.580	307	1162.121	407	1540.663
8	30.283	108	408.825	208	787.366	308	1165.907	408	1544.448
9	34.069	109	412.610	209	791.151	309	1169.692	409	1548.233
10	37.854	110	416.395	210	794.937	310	1173.478	410	1552.019
11	41.640	111	420.181	211	798.722	311	1177.263	411	1555.804
12	45.425	112	423.966	212	802.507	312	1181.049	412	1559.590
13	49.210	113	427.752	213	806.293	313	1184.834	413	1563.375
14	52.996	114	431.537	214	810.078	314	1188.619	414	1567.161
15	56.781	115	435.322	215	813.864	315	1192.405	415	1570.946
16	60.567	116	439.108	216	817.649	316	1196.190	416	1574.731
17	64.352	117	442.893	217	821.434	317	1199.976	417	1578.517
18	68.137	118	446.679	218	825.220	318	1203.761	418	1582.302
19	71.923	119	450.464	219	829.005	319	1207.546	419	1586.088
20	75.708	120	454.249	220	832.791	320	1211.332	420	1589.873
21	79.494	121	458.035	221	836.576	321	1215.117	421	1593.658
22	83.279	122	461.820	222	840.361	322	1218.903	422	1597.444
23	87.065	123	465.606	223	844.147	323	1222.688	423	1601.229
24	90.850	124	469.391	224	847.932	324	1226.473	424	1605.015
25	94.635	125	473.177	225	851.718	325	1230.259	425	1608.800
26	98.421	126	476.962	226	855.503	326	1234.044	426	1612.585
27	102.206	127	480.747	227	859.289	327	1237.830	427	1616.371
28	105.992	128	484.533	228	863.074	328	1241.615	428	1620.156
29	109.777	129	488.318	229	866.859	329	1245.401	429	1623.942
30	113.562	130	492.104	230	870.645	330	1249.186	430	1627.727
31	117.348	131	495.889	231	874.430	331	1252.971	431	1631.513
32	121.133	132	499.674	232	878.216	332	1256.757	432	1635.298
33	124.919	133	503.460	233	882.001	333	1260.542	433	1639.083
34	128.704	134	507.245	234	885.786	334	1264.328	434	1642.869
35	132.489	135	511.031	235	889.572	335	1268.113	435	1646.654
36	136.275	136	514.816	236	893.357	336	1271.898	436	1650.440
37	140.060	137	518.601	237	897.143	337	1275.684	437	1654.225
38	143.846	138	522.387	238	900.928	338	1279.469	438	1658.010
39	147.631	139	526.172	239	904.713	339	1283.255	439	1661.796

(CONTINUED)

VOLUME—CAPACITY (continued)
LIQUID OUNCES, QUARTS, AND GALLONS TO LITERS*

	Liters		Liters		Liters		Liters		Liters
40	151.417	140	529.958	240	908.499	340	1287.040	440	1665.581
41	155.202	141	533.743	241	912.284	341	1290.825	441	1669.367
42	158.987	142	537.529	242	916.070	342	1294.611	442	1673.152
43	162.773	143	541.314	243	919.855	343	1298.396	443	1676.937
44	166.558	144	545.099	244	923.641	344	1302.182	444	1680.723
45	170.344	145	548.885	245	927.426	345	1305.967	445	1684.508
46	174.129	146	552.670	246	931.211	346	1309.753	446	1688.294
47	177.914	147	556.456	247	934.997	347	1313.538	447	1692.079
48	181.700	148	560.241	248	938.782	348	1317.323	448	1695.865
49	185.485	149	564.026	249	942.568	349	1321.109	449	1699.650
50	189.271	150	567.812	250	946.353	350	1324.894	450	1703.435
51	193.056	151	571.597	251	950.138	351	1328.680	451	1707.221
52	196.841	152	575.383	252	953.924	352	1332.465	452	1711.006
53	200.627	153	579.168	253	957.709	353	1336.250	453	1714.792
54	204.412	154	582.953	254	961.495	354	1340.036	454	1718.577
55	208.198	155	586.739	255	965.280	355	1343.821	455	1722.362
56	211.983	156	590.524	256	969.065	356	1347.607	456	1726.148
57	215.769	157	594.310	257	972.851	357	1351.392	457	1729.933
58	219.554	158	598.095	258	976.636	358	1355.177	458	1733.719
59	223.339	159	601.881	259	980.422	359	1358.963	459	1737.504
60	227.125	160	605.666	260	984.207	360	1362.748	460	1741.289
61	230.910	161	609.451	261	987.993	361	1366.534	461	1745.075
62	234.696	162	613.237	262	991.778	362	1370.319	462	1748.860
63	238.481	163	617.022	263	995.563	363	1374.105	463	1752.646
64	242.266	164	620.808	264	999.349	364	1377.890	464	1756.431
65	246.052	165	624.593	265	1003.134	365	1381.675	465	1760.217
66	249.837	166	628.378	266	1006.920	366	1385.461	466	1764.002
67	253.623	167	632.164	267	1010.705	367	1389.246	467	1767.787
68	257.408	168	635.949	268	1014.490	368	1393.032	468	1771.573
69	261.193	169	639.735	269	1018.276	369	1396.817	469	1775.358
70	264.979	170	643.520	270	1022.061	370	1400.602	470	1779.144
71	268.764	171	647.305	271	1025.847	371	1404.388	471	1782.929
72	272.550	172	651.091	272	1029.632	372	1408.173	472	1786.714
73	276.335	173	654.876	273	1033.417	373	1411.959	473	1790.500
74	280.121	174	658.662	274	1037.203	374	1415.744	474	1794.285
75	283.906	175	662.447	275	1040.988	375	1419.529	475	1798.071
76	287.691	176	666.233	276	1044.774	376	1423.315	476	1801.856
77	291.477	177	670.018	277	1048.559	377	1427.100	477	1805.641
78	295.262	178	673.803	278	1052.345	378	1430.886	478	1809.427
79	299.048	179	677.589	279	1056.130	379	1434.671	479	1813.212
80	302.833	180	681.374	280	1059.915	380	1438.457	480	1816.998
81	306.618	181	685.160	281	1063.701	381	1442.242	481	1820.783
82	310.404	182	688.945	282	1067.486	382	1446.027	482	1824.569
83	314.189	183	692.730	283	1071.272	383	1449.813	483	1828.354
84	317.975	184	696.516	284	1075.057	384	1453.598	484	1832.139
85	321.760	185	700.301	285	1078.842	385	1457.384	485	1835.925
86	325.545	186	704.087	286	1082.628	386	1461.169	486	1839.710
87	329.331	187	707.872	287	1086.413	387	1464.954	487	1843.496
88	333.116	188	711.657	288	1090.199	388	1468.740	488	1847.281
89	336.902	189	715.443	289	1093.984	389	1472.525	489	1851.066
90	340.687	190	719.228	290	1097.769	390	1476.311	490	1854.852
91	344.473	191	723.014	291	1101.555	391	1480.096	491	1858.637
92	348.258	192	726.799	292	1105.340	392	1483.881	492	1862.423
93	352.043	193	730.585	293	1109.126	393	1487.667	493	1866.208
94	355.829	194	734.370	294	1112.911	394	1491.452	494	1869.993
95	359.614	195	738.155	295	1116.697	395	1495.238	495	1873.779
96	363.400	196	741.941	296	1120.482	396	1499.023	496	1877.564
97	367.185	197	745.726	297	1124.267	397	1502.809	497	1881.350
98	370.970	198	749.512	298	1128.053	398	1506.594	498	1885.135
99	374.756	199	753.297	299	1131.838	399	1510.379	499	1888.921

* Adapted from Units of Weight and Measure, National Bureau of Standards

DRILL SIZE DECIMAL EQUIVALENTS – METRIC AND INCH

NUMBER AND LETTER DRILLS

Drill No.	Frac	Deci
80		.0135
79		.0145
	1/64	.0156
78		.0160
77		.0180
76		.0200
75		.0210
74		.0225
73		.0240
72		.0250
71		.0260
70		.0280
69		.0292
68		.0310
	1/32	.0313
67		.0320
66		.0330
65		.0350
64		.0360
63		.0370
62		.0380
61		.0390
60		.0400
59		.0410
58		.0420
57		.0430
56		.0465
	3/64	.0469
55		.0520
54		.0550
53		.0595
	1/16	.0625
52		.0635
51		.0670
50		.0700
49		.0730
48		.0760
	5/64	.0781
47		.0785
46		.0810
45		.0820
44		.0860
43		.0890
42		.0935
	3/32	.0938
41		.0960
40		.0980
39		.0995
38		.1015
37		.1040
36		.1065
	7/64	.1094
35		.1100
34		.1110
33		.1130
32		.116
31		.120
	1/8	.125
30		.129
29		.136
28		.140
	9/64	.141
27		.144
26		.147
25		.150
24		.152
23		.154
	5/32	.156
22		.157
21		.159
20		.161
19		.166
18		.170
	11/64	.172
17		.173
16		.177
15		.180
14		.182
13		.185
	3/16	.188
12		.189
11		.191
10		.194
9		.196
8		.199
7		.201
	13/64	.203
6		.204
5		.206
4		.209
3		.213
	7/32	.219
2		.221
1		.228
A		.234
	15/64	.234
B		.238
C		.242
D		.246
	1/4	.250
E		.250
F		.257
G		.261
	17/64	.266
H		.266
I		.272
J		.277
K		.281
	9/32	.281
L		.290
M		.295
	19/64	.297
N		.302
	5/16	.313
O		.316
P		.323
	21/64	.328
Q		.332
R		.339
	11/32	.344
S		.348
T		.358
	23/64	.359
U		.368
	3/8	.375
V		.377
W		.386
	25/64	.391
X		.397
Y		.404
	13/32	.406
Z		.413
	27/64	.422
	7/16	.438
	29/64	.453
	15/32	.469
	31/64	.484
	1/2	.500
	33/64	.516
	17/32	.531
	35/64	.547
	9/16	.562
	37/64	.578
	19/32	.594
	39/64	.609
	5/8	.625
	41/64	.641
	21/32	.656
	43/64	.672
	11/16	.688
	45/64	.703
	23/32	.719
	47/64	.734
	3/4	.750
	49/64	.766
	25/32	.781
	51/64	.797
	13/16	.813
	53/64	.828
	27/32	.844
	55/64	.859
	7/8	.875
	57/64	.891
	29/32	.906
	59/64	.922
	15/16	.938
	61/64	.953
	31/32	.969
	63/64	.984
	1	1.000

METRIC DRILLS

MM	DEC.	MM	DEC.	MM	DEC.	MM	DEC.
1.	.0394	3.2	.1260	6.3	.2480	9.5	.3740
1.05	.0413	3.25	.1280	6.4	.2520	9.6	.3780
1.1	.0433	3.3	.1299	6.5	.2559	9.7	.3819
1.15	.0453	3.4	.1339	6.6	.2598	9.75	.3839
1.2	.0472	3.5	.1378	6.7	.2638	9.8	.3858
1.25	.0492	3.6	.1417	6.75	.2657	9.9	.3898
1.3	.0512	3.7	.1457	6.8	.2677	10.	.3937
1.35	.0531	3.75	.1476	6.9	.2717	10.5	.4134
1.4	.0551	3.8	.1496	7.	.2756	11.	.4331
1.45	.0571	3.9	.1535	7.1	.2795	11.5	.4528
1.5	.0591	4.	.1575	7.2	.2835	12.	.4724
1.55	.0610	4.1	.1614	7.25	.2854	12.5	.4921
1.6	.0630	4.2	.1654	7.3	.2874	13.	.5118
1.65	.0650	4.25	.1673	7.4	.2913	13.5	.5315
1.7	.0669	4.3	.1693	7.5	.2953	14.	.5512
1.75	.0689	4.4	.1732	7.6	.2992	14.5	.5709
1.8	.0709	4.5	.1772	7.7	.3031	15.	.5906
1.85	.0728	4.6	.1811	7.75	.3051	15.5	.6102
1.9	.0748	4.7	.1850	7.8	.3071	16.	.6299
1.95	.0768	4.75	.1870	7.9	.3110	16.5	.6496
2.	.0787	4.8	.1890	8.	.3150	17.	.6693
2.05	.0807	4.9	.1929	8.1	.3189	17.5	.6890
2.1	.0827	5.	.1968	8.2	.3228	18.	.7087
2.15	.0846	5.1	.2008	8.25	.3248	18.5	.7283
2.2	.0866	5.2	.2047	8.3	.3268	19.	.7480
2.25	.0886	5.25	.2067	8.4	.3307	19.5	.7677
2.3	.0906	5.3	.2087	8.5	.3346	20.	.7874
2.35	.0925	5.4	.2126	8.6	.3386	20.5	.8071
2.4	.0945	5.5	.2165	8.7	.3425	21.	.8268
2.45	.0965	5.6	.2205	8.75	.3445	21.5	.8465
2.5	.0984	5.7	.2244	8.8	.3465	22.	.8661
2.6	.1024	5.75	.2264	8.9	.3504	22.5	.8858
2.7	.1063	5.8	.2283	9.	.3543	23.	.9055
2.75	.1083	5.9	.2323	9.1	.3583	23.5	.9252
2.8	.1102	6.	.2362	9.2	.3622	24.	.9449
2.9	.1142	6.1	.2402	9.25	.3642	24.5	.9646
3.	.1181	6.2	.2441	9.3	.3661	25.	.9843
3.1	.1220	6.25	.2461	9.4	.3701		

TAP DRILL SIZES FOR UNIFIED STANDARD SCREW THREADS

Screw Thread Major Diameter	Threads Per Inch	Tap Drill Size Or Number	Screw Thread Major Diameter	Threads Per Inch	Tap Drill Size Or Number
0	80	3/64	3/8	16	5/16
				24	Q
1	64	53	7/16	14	U
	72	53		20	25/64
2	56	50	1/2	13	27/64
	64	50		20	29/64
3	48	47	9/16	12	31/64
	56	45		18	33/64
4	40	43	5/8	11	17/32
	48	42		18	37/64
5	40	38	3/4	10	21/32
	44	37		16	11/16
6	32	36	7/8	9	49/64
	40	33		14	13/16
8	32	29	1	8	7/8
	36	29		12	59/64
10	24	25	1 1/8	7	63/64
	32	21		12	1 3/64
12	24	16	1 1/4	7	1 7/64
	28	14		12	1 11/64
1/4	20	7	1 3/8	6	1 7/32
	28	3		12	1 19/64
5/16	18	F	1 1/2	6	1 11/32
	24	I		12	1 27/64

TAP DRILL SIZES FOR ISO METRIC THREADS

Nominal Size mm	Series Coarse Pitch mm	Tap Drill mm	Series Fine Pitch mm	Tap Drill mm
10	1.5	8.5	1.25	8.75
12	1.75	10.25	1.25	10.50
14	2	12.00	1.5	12.50
16	2	14.00	1.5	14.50
18	2.5	15.50	1.5	16.50
20	2.5	17.50	1.5	18.50
22	2.5	19.50	1.5	20.50
24	3	21.00	2	22.00
27	3	24.00	2	25.00

Nominal Size mm	Series Coarse Pitch mm	Tap Drill mm	Series Fine Pitch mm	Tap Drill mm
1.4	0.3	1.1	—	—
1.6	0.35	1.25	—	—
2	0.4	1.6	—	—
2.5	0.45	2.05	—	—
3	0.5	2.5	—	—
4	0.7	3.3	—	—
5	0.8	4.2	—	—
6	1.0	5.0	—	—
8	1.25	6.75	1	7.0

EXERCISE 1-1

2. Eighty-three
4. Five hundred ninety-seven
6. One hundred five thousand, seven hundred seventy-seven
8. One hundred thousand, one
10. Six hundred eighty-two million, fifty-one thousand, four hundred thirty-nine
12. 73
14. 200
16. 23,422
18. 947,682
20. 978,633
22. 3,214,006

EXERCISE 2-1

2. 100
4. a. thousandths
   b. hundreths
   c. tenths
   d. tens
   e. hundreds
   f. thousands
6. a. centimeter
   b. centimeter, meter
8. two
10. liters
12. 1000
14. a. (local answers vary)
    b. (local answers vary)
    c. H = 280 millimeters
       W = 215 millimeters
       T = 8 millimeters
          (approximately)
16. a. 3.785 liters
    b. (answers vary)

EXERCISE 3-1

2. 1501
4. 880

6. 12,519
8. 22,466
10. 19,485
12. 30,258
14. 34,466

EXERCISE 3-2

2. 3354 lb.
4. 1455 cm

EXERCISE 4-1

2. 9
4. 732
6. 478
8. 143
10. 101
12. 1660
14. 4625
16. 3857
18. 62,969
20. 92,814,000
22. 0
24. 18,042
26. 1
28. 36,642
30. 7387
32. 238,945

EXERCISE 4-2

2. 1218 feet
4. 767 feet
6. 850 pounds
8. 96
10. 241
12. 1525 kilometers
14. 2154

EXERCISE 5-1

2. 575
4. 3999
6. 2117

8. 161,240
10. 3,304,909
12. 6,142,500
14. 343,266
16. 43,142
18. 247,434
20. 169,347,336

EXERCISE 5-2

2. 70 seconds or 1 minute 10 seconds
4. 1968 kilograms
6. 2,703,360
8. 768 candles
10. 308 kilograms

EXERCISE 6-1

2. 54
4. 45
6. 18
8. 1620 R2
10. 996
12. 5060 R3
14. 423 R37
16. 1052 R31

EXERCISE 6-2

2. 4220
4. 187
6. 92
8. $26,400
10. 50
12. 46

EXERCISE 7-1

2. 320/8
4. 81/3
6. 650/5

EXERCISE 7-2

2. 35/4

4. 235/32
6. 321/64

EXERCISE 7-3

2. 11
4. 4 7/9
6. 31 5/6

EXERCISE 7-4

2. 1/3
4. 3/8
6. 1/2

EXERCISE 7-5

2. 6/32
4. 21/49
6. 93/237

EXERCISE 8-1

2. 1/7
4. 1 4/9
6. 1 1/8
8. 3
10. 1 3/8
12. 6 3/8
14. 26
16. 1 3/16
18. 7/12
20. 12 1/2
22. 137 14/15
24. 473 13/27
26. 19 49/90
28. 31 39/56
30. 76 11/16

EXERCISE 8-2

2. 3/4 inch
4. 122 7/8 inches
6. 81 3/4 inches
8. 10 5/8 inches
10. 131 3/8 inches

EXERCISE 9-1

2. 9/16
4. 1 5/8
6. 15 37/90
8. 56 23/42
10. 63 33/68
12. 12 27/32
14. 10 3/5
16. 83 1/12
18. 3/16

20. 27 5/13
22. 205 23/40
24. 5060 4/9
26. 4777

EXERCISE 9-2

2. 4922 1/2 pounds
4. 16/1000 inch or 2/125
6. 1 1/2 inches
8. 1 5/32
10. 5 5/16 inches

EXERCISE 10-1

2. 1/4
4. 10/33
6. 3/128
8. 21/32
10. 5/72
12. 15/256
14. 1
16. 1 3/4
18. 1
20. 18 2/3
22. 1/6
24. 25 2/3

EXERCISE 10-2

2. 15/32 inch
4. 3/4 inch
6. 3/32 ounces
8. 24 3/4 inches

EXERCISE 10-3

2. 22 1/2
4. 42 19/24
6. 8
8. 1
10. 130 11/48
12. 50 25/72
14. 13 3/4
16. 10 20/21
18. 1 5/6
20. 11

EXERCISE 10-4

2. 1375 inches
4. 4687 1/2 inches
6. 693 3/4 inches
8. 7 1/3 cubic yards
10. 47 inches
12. 114 3/8 inches or 9 feet
6 3/8 inches

14. 58 3/16 pounds
16. 52 1/2 meters

EXERCISE 11-1

2. 1
4. 5
6. 144
8. 10 1/2
10. 2 2/3
12. 8 13/14
14. 2/5
16. 9/70
18. 5/12
20. 1 15/28
22. 2 16/17
24. 2
26. 25/38
28. 1 17/26
30. 10 7/8

EXERCISE 11-2

2. 12
4. 22 pieces and 4 inches left
6. 4 pieces and none left
8. 7 7/8
10. 37 1/2 minutes
12. 12

EXERCISE 12-1

2. .075
4. .46
6. 2.4
8. .72
10. 93.0017
12. .34
14. .097
16. .0008
18. 163.091
20. 911.3

EXERCISE 12-2

2. .24
4. .0035
6. .387
8. .203
10. .009
12. .02050
14. 24.603
16. 5.3112
18. 23.0006
20. .475
22. .054

24. 79.3
26. .0268
28. 1001.0001
30. .0097

EXERCISE 12-3

2. .27
4. 117.0711
6. 1016.3
8. .099
10. .1025
12. .2050
14. .0014
16. .096
18. .0172
20. 119.91
22. 23.9
24. .0006

EXERCISE 12-4

2. .75
4. 375/1000 or 3/8
6. 145/1000 or 29/200
8. 91/1000
10. 948/1000 or 237/250
12. .0023
14. 2 993/1000

EXERCISE 13-1

2. 1.072
4. .0102
6. 4.508
8. 13.132
10. 5.153
12. 117.21
14. 7.1
16. 36.2
18. .4000
20. 6.455
22. 180.99
24. 845.75
26. 6022.60
28. $38.31
30. 3299.88
32. 1405.0068
34. $695.99
36. 56.000

EXERCISE 13-2

2. 141.50 feet
4. $1491.48
6. 5.753 inches

8. 38.5 inches
10. 722.375 feet
12. 44.18 centimeters
14. 32.05 centimeters
16. 36.73 centimeters

EXERCISE 14-1

2. .50
4. 11.1
6. 5.9
8. 5.7
10. 5.7
12. 3.91
14. 5.62
16. 6.03
18. 6.87
20. 3.786
22. 22.899
24. 29.2363
26. 24.4637
28. .35010
30. 800.610
32. 89.586
34. .125000
36. .515625
38. 39.6821
40. 11.6914
42. .00518
44. 1.9324
46. .1803
48. .107
50. 19.48
52. 0
54. 7.000

EXERCISE 14-2

2. 78.875 centimeters
4. $113.10
6. 125.25 pounds

EXERCISE 15-1

2. 1.4427
4. .0385
6. 42.12
8. .194
10. 5.56
12. .3376
14. 23.5
16. 259,000
18. .7524
20. 112.064
22. 11,889.29
24. .0008416

26. 2.011015

EXERCISE 15-2

2. 199.424 square inches
4. 6.53125 m²
6. $45.08

EXERCISE 16-1

2. 3
4. .2
6. 7
8. 3
10. 1.36
12. .864
14. .5
16. 3.69
18. 36.9
20. 7500
22. 2.4
24. .0003
26. 29.31

EXERCISE 16-2

2. 9
4. 27
6. 21
8. 7.857
10. 60.12
12. .586 pounds

EXERCISE 17-1

2. .031
4. .063
6. .094
8. .156
10. .25
12. .344
14. .391
16. .484
18. .578
20. .797
22. .922

EXERCISE 17-2

2. 157.5 inches
4. 2.1875 inches
6. 8.75 inches
8. 1 3/16 = 1.188;
   1 3/8 = 1.375;
   1 7/32 = 1.219;
      11/16 = .688;
   3 1/8 = 3.125;

$$7/16 = .438;$$
$$6 \ 5/16 = 6.313;$$
$$4 \ 7/16 = 4.438;$$
$$9 \quad = 9.0 \ ;$$
$$15/16 = .938;$$
$$7/8 = .875;$$
$$3/8 = .375;$$

EXERCISE 18-1

2. 5/8
4. 3/32
6. 1/2
8. 9/16
10. 5/32
12. 5/16
14. 11/16

EXERCISE 19-1

2. 879.348 millimeters
4. 14.6304 meters
6. 43.4511 kilometers
8. .5713 inches
10. 60.8972 miles
12. 16.8 m²
14. 3.6 Mg
16. 22.938 m³
18. 11.352 ℓ
20. 4.24 quarts
22. 3263.9 millimeters
24. 30.94 m²
26. 3.785 ℓ
28. 3.783 meters
30. 2 ℓ

EXERCISE 19-2

2. 6.7056 meters × 9.7536
   = 65.507641 m²
4. 2937.6 kilograms;
   2.937.6 Mg
6. 76.113408 m³
8. 88.9475
10. 77 °F

EXERCISE 20-1

2. 24.05
4. 4.05
6. 30
8. 106.4
10. 724.2
12. 148.2
14. 122.88
16. 500

18. 1
20. 200
22. 2475
24. 630
26. 57.4
28. 8.96
30. 33.32
32. 30
34. 6.12
36. 68.95
38. .07

EXERCISE 20-2

2. $228.48
4. $2418.00
6. $1462.50
8. 5%
10. 6
12. .025%
14. $19.13
16. 225

EXERCISE 21-1

2. $41.28
4. $43.88
6. $70.93
8. $96.60
10. $2.48
12. $2.50
14. $2.07
16. $1.60
18. $12.15
20. $21.01
22. $1500
24. $1.30

EXERCISE 22

Answers are found in Unit 22

EXERCISE 23-1

2. 8
4. 10
6. 40
8. 7/8
10. 1 1/2
12. 26
14. 42
16. 1 7/8

EXERCISE 23-2

b. 3/4
d. 2 1/4

f. 2 5/16
h. 5 1/2
j. 4 9/16
l. 1 3/8
n. 4 11/16
p. 4 7/8
r. 4 7/16
t. 3 1/4

EXERCISE 23-3

b. 4 3/4
d. 1/4
f. 3/4
h. 1 9/16
j. 5/8
l. 1 1/2
n. 9/16

EXERCISE 23-4

b. 5/8
d. 1 1/4
f. 5/16
h. 1/4
j. 7/16
l. 3/4
n. 1 3/8
p. 3 3/8

EXERCISE 23-5

B. .40
D. .75
F. 1.24
H. 1.71

EXERCISE 23-6

B. 23 millimeters
D. 47 millimeters
F. 74 millimeters
H. 100 millimeters

EXERCISE 24-1

b. .235
d. .375
f. .250
h. .468
j. .718

EXERCISE 24-2

Micrometer readings are given and are to be drawn in by the student.

EXERCISE 24-3

b. 8.15 millimeters
d. 15.75 millimeters
f. 3.05 millimeters
h. 5.55 millimeters
j. 20.08 millimeters